主动控制术语汉英对照手册

Chinese-English manual for active control terminology

王 飞 吴文伟 苏常伟 主 编

哈尔滨工程大学出版社
Harbin Engineering University Press

内 容 简 介

振动噪声的主动控制方兴未艾,以新材料、新方法和深度学习为代表的人工智能技术的快速发展更是大大加快了其发展步伐。本书的作者深感普及主动控制基本理论、基本概念等的紧迫性,以数百篇文献为基础,编纂了这部有关主动控制常用术语的英汉对照手册。

主动控制是一个跨领域、多学科融合的综合性专业,虽然其所涉及的各专业知识在自己所属领域可能不是新兴事物,但是从主动控制的角度,却提出了新的要求或引申了新的含义。本书从主动控制专业研究领域的角度出发,总结提炼该领域关键词汇。本书对于认识、理解与掌握主动控制具有重要意义。

本书可供对主动控制感兴趣的在校学生、一线工程人员及其他专业人员参考阅读。

图书在版编目(CIP)数据

主动控制术语汉英对照手册/王飞,吴文伟,苏常伟主编. —哈尔滨:哈尔滨工程大学出版社,2020.11
ISBN 978 – 7 – 5661 – 2867 – 6

Ⅰ.①主… Ⅱ.①王…②吴…③苏… Ⅲ.①振动控制—名词术语—手册—汉、英②噪声控制—名词术语—手册—汉、英 Ⅳ.①TB535 – 62

中国版本图书馆 CIP 数据核字(2020)第 228723 号

主动控制术语汉英对照手册
ZHUDONG KONGZHI SHUYU HANYING DUIZHAO SHOUCE

选题策划	史大伟　薛　力
责任编辑	薛　力
封面设计	李海波

出版发行	哈尔滨工程大学出版社
社　　址	哈尔滨市南岗区南通大街 145 号
邮政编码	150001
发行电话	0451 – 82519328
传　　真	0451 – 82519699
经　　销	新华书店
印　　刷	北京中石油彩色印刷有限责任公司
开　　本	787 mm × 1 092 mm　1/16
印　　张	7
字　　数	183 千字
版　　次	2020 年 11 月第 1 版
印　　次	2020 年 11 月第 1 次印刷
定　　价	40.00 元

http://www.hrbeupress.com
E-mail:heupress@ hrbeu.edu.cn

前　　言

自第一项有关主动控制的专利申请以来，主动控制已经发展了将近90年，其间几次起起伏伏，现在基于微电子技术、传感技术、智能材料以及控制理论的巨大进步，主动控制也进入了蓬勃发展阶段。

主动控制现已成功应用于航天、航空、轨道交通、汽车、家用电器、数码产品、建筑物等多个领域，在延长设备使用寿命、保障运行精度、提升驾乘体验、提高生活品质等方面起着重要作用。

主动控制涉及力学、声学、自动控制、计算机、电子工程、软件工程等多个学科，对希望通过主动控制手段减振降噪的人来说是一个典型的跨专业、跨学科的全新领域。因此，阅读相关文献，以此吸取前人知识和经验就变得非常必要。

本书作者作为振动噪声领域的从业人员，在科研攻关中以主动控制解决了传统控制手段所不能解决的振动噪声问题，深刻感受到主动控制必将取代被动控制成为未来研究的热点。

另外，有感于当初入门学习主动控制时的困难和挫折，特以数百篇经典主动控制文献为基础，编纂这本术语手册，以方便后来者。

鉴于作者能力有限，书中如有疏漏或错误还请批评指正。

编　者

2020 年 10 月

目 录

1 振动 ··· 1
　1.1 振动与噪声产生的原因 ·· 1
　1.2 振动与噪声的危害与利用 ·· 1
　1.3 振动与噪声的控制 ·· 1
2 数学 ··· 3
　2.1 符号 ··· 3
　2.2 单位 ··· 4
　2.3 高等数学 ·· 5
　2.4 矩阵 ··· 9
　2.5 概率与数理统计 ·· 11
3 力学 ··· 15
　3.1 运动 ··· 15
　3.2 应力、应变与材料特性 ·· 15
　3.3 固体力学 ·· 16
　3.4 振动 ··· 16
4 声学 ··· 18
　4.1 基本概念 ·· 18
　4.2 声测量 ··· 19
　4.3 阻抗 ··· 19
　4.4 流体动力学 ··· 19
　4.5 声辐射 ··· 20
5 自动控制 ··· 21
　5.1 基本概念 ·· 21
　5.2 模型 ··· 22
　5.3 系统分析 ·· 23
　5.4 离散系统 ·· 24
　5.5 现代控制理论 ··· 24
6 电力电子 ··· 26
　6.1 基本概念 ·· 26
　6.2 电路计算 ·· 26
　6.3 电子 ··· 27
　6.4 运算电路 ·· 28
　6.5 电力 ··· 28

7 传感器和作动器 29
7.1 基本概念 29
7.2 传感器 29
7.3 作动器 31
7.4 压电材料 32
7.5 其他材料 32

8 仿真计算 34
8.1 数字信号处理 34
8.2 控制算法 34
8.3 仿真软件 36

9 系统搭建 38
9.1 机械加工 38
9.2 装配 39
9.3 仪器设备 39

10 应用领域 41
10.1 通用 41
10.2 航空航天 42
10.3 车辆 42
10.4 舰船 43
10.5 建筑物 43
10.6 家用电器、消费类电子 43

附录 A 英汉对照 45

附录 B 核对与检查计算公式和结果 77

附录 C 英语论文写作常用词汇 78
C.1 连接词汇 78
C.2 常用动词及动词短语 81
C.3 常用副词 84
C.4 讨论 85
C.5 图表 87

参考文献 90

1 振 动

1.1 振动与噪声产生的原因

不平衡的 out-of-balance
往复的 reciprocating
摄动 perturbation

1.2 振动与噪声的危害与利用

平民 civilian
鼓膜 eardrum
可听阈 threshold of audibility
听力学 audiology
听力损害 hearing loss
  暂时的 temporary
  永久的 abiding
标准的 canonical
有害的 deleterious
恶化,变坏 deteriorate
损害,伤害 detriment
有扩散危害的 invasive
能量收集 energy harvest

1.3 振动与噪声的控制

1.3.1 传统控制手段

隔离 isolation
整体密闭罩 integral enclosure

1.3.2 传统控制手段的不足

(巨大)物体 bulk
沉重的 cumbersome
大而重的 massive
失去作用 played out

1.3.3 主动控制

顺应性 compliance
学科 discipline
各学科间的 interdisciplinary
(某一学科的)术语,专门名称 nomenclature
施加 impose
轻量的 lightweight
小型的 miniature
使小型化 miniaturise
使小型化 miniaturization
成熟 maturity
偏离共振 off-resonant
反相 out of phase
相移 phase shift
物理观点 physical insight
物理原理 physical principle
对抗 react off
压制 suppress
对…起作用 react on
结构噪声 structure borne noise
共振项 resonance term
噪声消除 noise cancellation
抵消 cancel out
对抗,抵消 counteract
对消点 cancellation point
相干频率 coincidence frequency
紧凑的 compact
互补的 complementary
有成本效益的 cost-effective
方向性 directivity

差异 discrepancy
下游 downstream
激励 excitation
激励频率 excitation frequency
扰动频率 forcing frequency
同步 in-phase
相互作用 interaction
互惠 reciprocity
拦截 intercept
可交换地 interchangeably
互相连接 interconnect
暂时的 interim
间歇的 intermittent
次级的 secondary

2 数　　学

2.1 符　　号

符号规约 sign convention

常用的符号如下：

符号	中文表达	英文表达
+	加；正	plus
-	减；负	minus
±	正负	plus or minus
×	乘	is multiplied by
÷	除	is divided by
=	等于	is equal to
≠	不等于	is not equal to
≡	全等于	is equivalent to
≅	等于或约等于	is equal to or approximately equal to
≈	约等于	is approximately equal to
<	小于	is less than
>	大于	is more than
≮	不小于	is not less than
≯	不大于号	is not more than
≤	小于或等于号	is less than or equal to
≥	大于或等于号	is more than or equal to
%	百分之	per cent
‰	千分之	per mill
∞	无限大号	infinity
∝	与…成比例	varies as
√	平方根	(square) root

(续)

符号	中文表达	英文表达
∵	因为	since, because
∴	所以	hence
∷	等于,成比例	equals, as (proportion)
∠	角	angle
⌒	半圆	semicircle
⊙	圆	circle
○	圆周	circumference
π	圆周率	pi
△	三角形	triangle
⊥	垂直于	perpendicular to
∪	并,合集	union of
∩	交,通集	intersection of
∫	…的积分	the integral of
Σ	总和	(sigma) summation of
°	度	degree
′	分	minute
″	秒	second
℃	摄氏度	Celsius system
{	左花括号	open brace, open curly
}	右花括号	close brace, close curly
(	左圆括号	open parenthesis, open paren
)	右圆括号	close parenthesis, close paren
( )	括号	brackets/parentheses
[	左方括号	open bracket
]	右方括号	close bracket
[ ]	方括号	square brackets
.	句号,点	period, dot, full stop
		竖线	vertical bar, vertical virgule
&	和,引用	ampersand, and, reference, ref

(续)

符号	中文表达	英文表达
*	星号,乘号,星,指针	asterisk, multiply, star, pointer
/	斜线,斜杠,除号	slash, divide, oblique
//	双斜线,注释符	slash-slash, comment
#	井号	pound
\	反斜线转义符,有时表示转义符或续行符	backslash, sometimes escape
~	波浪符	tilde
,	逗号	comma
:	冒号	colon
;	分号	semicolon
?	问号	question mark
!	感叹号	exclamation mark（英式英语）exclamation point（美式英语）
'	撇号	apostrophe
-	连字号	hyphen
——	破折号	dash
..	省略号	dots/ ellipsis
"	单引号	single quotation marks
" "	双引号	double quotation marks
∥	双线号	parallel
&	与	ampersand, and
~	或,代字号	swung dash
§	分节号	section, division
→	箭号	arrow

四则运算 Arithmetic
 加 addition
 减 subtraction
 乘 multiply
 除 divide
 和 sum
 差 difference
 商 quotient
 积 product
分式 fraction
 分子 numerator
 分母 denominator
克罗内克符号 Kronecker symbol
记号表达式 notational expression
差数 margin
算子记号 operator notation
运算符 operator
表音符号 phonogram
下标的 subscript
上标 superscript

单位的十进制倍数：

1 000 000 000 000 ×	10^{12}	Tera-	T
1 000 000 000 ×	10^{9}	Giga-	G
1 000 000 ×	10^{6}	Mega-	M
1 000 ×	10^{3}	Kilo-	k
100 ×	10^{2}	Hector-	h
10 ×	10^{1}	Deka-	da
0.1 ×	10^{-1}	Deci-	d
0.01 ×	10^{-2}	Centi-	c
0.001 ×	10^{-3}	Milli-	m
0.000 001 ×	10^{-6}	Micro-	μ
0.000 000 001 ×	10^{-9}	Nano-	n
0.000 000 000 001 ×	10^{-12}	Pico-	p

2.2 单 位

国际单位制, System International of Units, 简称 SI, 国际计量大会（General Conference of Weights & Measures-CGPM）采纳和推荐的一种一贯单位制。在国际单

位制中,将单位分成三类:基本单位(base units)、导出单位(derived units)和辅助单位。

辅助单位目前只有两个,纯系几何单位,分别是:

弧度(rad)是一个圆内两条半径在圆周上截取的弧长与半径相等时,它们所夹的平面角的大小。

球面角(sr)是一个立体角,其顶点位于球心,而它在球面上所截取的面积等于以球半径为边长的正方形面积。

当然,辅助单位也可以进一步再构成导出单位。

一致性量纲 consistent units

无量纲的 dimensionless, non-dimensional

2.2.1 基本单位

7个严格定义的基本单位是:

中文名称	英文名称	物理符号	中文名称	单位符号
长度	Length	L	米	m
质量	Mass	m	千克	kg
时间	Time	t	秒	s
电流	Current	I	安培	A
热力学温度	Temperature	T	开尔文	k
物质的量	Mole	$n(v)$	摩尔	mol
发光强度	Luminous intensity	$I(Iv)$	坎德拉	cd

基本单位在量纲上彼此独立。

2.2.2 导出单位

导出单位很多,都是由基本单位组合起来而构成的。

常用的导出单位有:

中文名称	英文名称	物理符号	中文名称	单位符号
面积	Area	$A(S)$	平方米	m^2
体积	Volume	V	立方米	m^3
速度	Velocity	v	米每秒	m/s
加速度	Acceleration	a	米每秒方	m/s^2
角加速度	Angular Acceleration	ω	弧度每秒	rad/s
力	Force	f	牛顿	$kg \cdot m/s^2 = N$
能量	Energy	E	焦耳	$kg \cdot m^2/s^2 = J$
功率	Power	W	瓦特	$kg \cdot m^2/s^3 = W$
密度	Density	ρ	千克每立方米	kg/m^3
电量	Coulomb	Q	库仑	$A \cdot s = C$
电压	Voltage	V	伏特	$V = W/A$
电容	Capacitance	C	法拉	$F = C/V$
磁通	Magnetic Flux	Φ	韦伯	$Wb = V \cdot s$
电感	Inductance	L	亨利	$H = Wb/A$

2.3 高等数学

2.3.1 数

数 number

整数 integer

小数 decimal

自然数 natural number

对数 logarithm

有理数 rational number

无理数 irrational number

实数 real number

虚数 imaginary number

十进制的 decimal

八进制的 octal

二进制的 binary

十六进制的 hexadecimal

引理 lemma

定理 theorem

公理 axiom

整数倍数 integer multiple

最低有效位 least significant bit(LSB)

最高有效位 most significant bit(MSB)

零 nought

零值的 null

代数的 algebraical

2.3.2 微积分

微积分 calculus

线性近似 linear approximation

封闭区间 closed interval

开区间 open interval

增量 increment

乘子 multiplier

定积分 definite integral

不定积分 indefinite integral

瑕积分 improper integral

围道积分 contour integral

近似积分 approximate integration

三角积分 trigonometric integrals

三角代换法 trigonometric substitutions

二重的 dual

二元性 duality

二重积分 double integral

分部积分法 integration by part

三重积分 tripe integrals

多重积分 multiple integrals

体积 volume

曲面 surface

面积分 surface integral

旋转曲面 surface of revolution

部分分式 partial fractions

部分积分 partial integration

导数 derivative

高阶 higher-order

全微分 total differential

常微分 ordinary differential

偏导数 partial derivative

右导数 right-hand derivative

左导数 left-hand derivative

右极限 right-hand limit

左极限 left-hand limit

鞍点 saddle point

偏微分方程 partial differential equation

参数 parameter

奇点 singularity

可导函数 differentiable function

不连续性 discontinuity

梯度 gradient

反导数 anti-derivative

一阶导数试验法 first derivative test

二阶导数 second derivative

二阶导数试验法 second derivative test

二阶偏导数 second partial derivative

方向导数 directional derivative

高阶导数 higher derivative

2.3.3 坐标系

坐标轴 axes

主轴 principal axes

坐标 coordinate

笛卡儿坐标 Cartesian coordinates

笛卡儿坐标系 Cartesian coordinate

system
    直角坐标 rectangular coordinates
    直角坐标系 rectangular coordinate system
    坐标轴 coordinate axes
    坐标平面 coordinate planes
    柱面坐标 cylindrical coordinate
    极坐标 polar coordinate
    球面坐标 spherical coordinate
    极轴 polar axis
    象限 quadrant
    卦限 octant
    第一卦限 first octant
    原点 origin
    正交的 orthogonal
    x 轴 x-axis
    x 坐标 x-coordinate
    截距 intercepts
    x 截距 x-intercept
    法线 normal line
    法向量 normal vector
    渐近线 asymptote
    垂直渐近线 vertical asymptote
    水平渐近线 horizontal asymptote
    斜渐近线 slant
    分割 partition
    割线 secant line
    斜渐近线 slant asymptote
    斜率 slope
    垂直线 perpendicular lines
    正角 positive angle
    切线 tangent line
    变化率 rate of change
    直线的斜截式 slope-intercept equation of a line
    点斜式 point-slope form
    切平面 tangent plane
    切向量 tangent vector

2.3.4 几何

    点 dot
    线 line
    曲线 curve
    螺旋线 helix
    平面 plane
    矩形 rectangle
    正方形 square
    三角形 triangle
    菱形 diamond
    梯形 trapezoid
    椭圆 ellipse
    圆 circle
    四面体 tetrahedron
    圆锥 cone
    圆锥形的 conical
    圆柱 cylinder
    柱 pillar
    球体 sphere
    椭圆体 ellipsoid
    壳 shell
    拐点 inflection point
    平滑曲线 smooth curve
    平滑曲面 smooth surface
    水平线 horizontal line
    双曲线 hyperbola
    双曲面 hyperboloid
    外摆线 epicycloid
    旋转体 solid of revolution
    旋转曲面 surface of revolution
    抛物线 parabola
    抛物柱面 parabolic cylinder
    抛物面 paraboloid
    平行六面体 parallelepiped
    并行线 parallel lines
    环形的 circular
    圆周的 circumferential

凹的 concave
凸面的 convex
六面体的 hexahedral
双曲线的 hyperbolic
等大的 isometric
等腰的 isosceles
等面的 equilateral
梯形的 trapezoidal
成三角形地 triangularly
三轴的 triaxial
垂直地 vertically
四分之一圆 quadrant
四边(形)的 quadrilateral
四倍的,四重的 quadruple
直角 right angle
球状地 spherically
截线 transversal
曲率半径 radius of curvature
线性尺寸 linear dimension

2.3.5 数列

级数 series
　　二项级数 binomial series
　　调和级数 harmonic series
　　几何级数 geometric series
　　无穷级数 infinite series
严格递减 strictly decreasing
严格递增 strictly increasing
递减函数 decreasing function
递减数列 decreasing sequence
收敛半径 radius of convergence
绝对收敛 absolute convergence
收敛 convergence
收敛区间 interval of convergence
收敛半径 radius of convergence
收敛数列 convergent sequence
收敛级数 convergent series
黎曼和 riemann sum

黎曼几何 riemannian geometry
整合序列 integration sequence
非递归的 non-recursive
非平稳 non-stationary
递归 recursion
递归的 recursive
最小上界 least upper bound
左边序列 left-sided sequence

2.3.6 函数

2.3.6.1 基本概念

函数值 value of function
变量 variable
自变量 independent variable
因变量 dependent variable
系数 coefficient
多项式的零点 zeros of a polynomial
根 root
夹挤定理 squeeze theorem
均值定理 mean value theorem
极限的商定律 quotient law of limit
商定律 quotient rule
中值定理 intermediate value of theorem
柯西均值定理 Cauch's mean value theorem
初值定理 initial-value theorem
非齐(次,性)的 nonhomogeneous
周期图 periodogram

2.3.6.2 连续性

在一点处之连续性 continuity at a point
函数之连续性 continuity of a function
在区间之连续性 continuity on an interval
区间 interval
左连续 continuity from the left

右连续 continuity from the right
连续函数 continuous function
线性方程式 linear equation
线性函数 linear function
线性 linearity
线性化 linearization
左极限 left-hand limit
有理代换法 rationalizing substitution
极值 extreme value
极值定理 extreme value theorem
相对极大值与极小值 relative maximum and minimum values
绝对极大与极小 absolute maximum and minimum
极大与极小值 maximum and minimum values
绝对极值 absolute extreme values
绝对值 absolute value
绝对值函数 absolute value function
局部极值 local extreme values
局部极大值与极小值 local maximum and minimum values
函数的值域 range of a function
周期 period
非周期 aperiodic
周期函数 periodic function
单边极限 one-sided limit
最优化问题 optimization problems
方程式 equation
期望值 expected valued
指数函数 exponential function
阶乘 factorial
分段定义函数 piecewise defined function
临界点 critical point
三次函数 cubic function
极方程式 polar equation
极点 pole

多项式 polynomial
系数 coefficient
多项式之次数 degree of a polynomial
反函数 inverse function
反三角函数 inverse trigonometric function

2.3.6.3 函数类型

奇函数 odd function
偶函数 even function
常数函数 constant function
对数函数 logarithmic function
隐函数 implicit function
有理函数 rational function
正弦函数 sine function
余弦函数 cosine function
正切函数 tangent function
三角函数 trigonometric function
自然指数函数 natural exponential function
自然对数函数 natural logarithm function
幂函数 power function
收入函数 revenue function
阶梯函数 step function
凸函数 convex function
凹函数 concave function
正弦曲线的 sinusoidal

2.4 矩 阵

向量 vector
标量 scalar
常量 constant
Jordon 标准型 Jordon canonical form
QR 分解 QR decomposition
Smith 标准型 Smith normal form
半正定 semi-positive definite

标准正交基 orthonormal basis
不变因子 invariant factor
充分必要条件 necessary and sufficient condition
初等因子 elementary divisor
纯虚数 pure imaginary number
单位下三角矩阵 unitary low triangular
对角化 diagonalizable
对角矩阵 diagonal matrix
反 hermite 矩阵 skew hermite matrix
反对称矩阵 anti-symmetric matrix/skew-symmetric matrix
范数 norm
长度 length
化零多项式 annihilating polynomial
基 base
极小范数解 minimum norm solution
极小最小二乘解 minimum least-squares solution
假设 hypothesis
矩阵 matrix
可对角化 diagonalizable
可逆 invertible
满秩分解 full-rank decomposition
幂级数 power series
内积空间 inner product spaces
逆矩阵 inverse matrix
谱 spectrum
奇异值 singular value
任意多项式 arbitrary polynomial
三角分解 triangle decomposition
上三角矩阵 upper triangular matrix
实对称矩阵 real symmetric matrix
收敛的 converged
收敛性 convergence
特征多项式 characteristic polynomial
特征向量 eigenvector
特征值 eigenvalue

通解 general solution
维数 dimension
线性变换 linear transform
线性方程组 linear equations
线性空间 linear space
线性无关 linear independence
线性相关 linear dependence
非共线矢量 non-collinear vectors
标准方程式 normal equations
非奇异的 non-singular
线性代数 linear algebra
标准化的 normalized
正定性 positive definiteness
不相关的 uncorrelated
自伴的 self-adjoint
单位阵 identity matrix
同一性 identity
病态 ill-conditioned
不适定问题 ill-posed problem
内积 inner product
外积 outer product
非对角元 off-diagonal element
转置阵 transpose
二次判别函数 quadratically integrable function
线性映射 linear mapping
相容范数 consistent norm/compatible norm
相容方程组 compatible equations
相似矩阵 similar matrix
向量空间 vector space
行列式 determinant
行列式因子 determinant factor
酉矩阵 unitary matrix/u-matrix
酉空间 unitary space
正规矩阵 normal matrix
正交的 orthogonal
值域 value range

重根 multiple roots
最小多项式 minimum polynomial
最小二乘解 least-squares solution
共轭的 conjugate
协方差 covariance

2.5 概率与数理统计

2.5.1 基本概念

集合 set
空间 space
不确定性 indeterminacy
必然现象 certain phenomenon
随机现象 random phenomenon
试验 experiment
结果 outcome
频率数 frequency number
样本空间 sample space
出现次数 frequency of occurrence
n 维样本空间 n-dimensional sample space
样本空间的点 point in sample space
随机事件 random event/random occurrence
基本事件 elementary event
必然事件 certain event
不可能事件 impossible event
等可能事件 equally likely event
事件运算律 operational rules of events
事件的包含 implication of events
并事件 union events
交事件 intersection events
互不相容事件，互斥事件 mutually exclusive events/incompatible events
互逆的 mutually inverse
加法定理 addition theorem
古典概率 classical probability
古典概率模型 classical probabilistic model
几何概率 geometric probability
乘法定理 product theorem
概率乘法 multiplication of probabilities
条件概率 conditional probability
全概率公式、全概率定理 formula of total probability
贝叶斯公式、逆概率公式 bayes formula
后验概率 posterior probability
先验概率 prior probability
独立事件 independent event
独立随机事件 independent random event
独立实验 independent experiment
两两独立 pairwise independent
两两独立事件 pairwise independent events

2.5.2 随机变量及其分布 (Random Variables and Distributions)

随机变量 random variables
离散随机变量 discrete random variables
概率分布律 law of probability distribution
一维概率分布 one-dimension probability distribution
概率分布 probability distribution
两点分布 two-point distribution
伯努利分布 Bernoulli distribution
二项分布/伯努利分布 Binomial distribution
超几何分布 hypergeometric distribution
三项分布 trinomial distribution
多项分布 polynomial distribution

泊松分布 Poisson distribution
泊松定理 Poisson theorem
分布函数 distribution function
概率分布函数 probability distribution function
连续随机变量 continuous random variable
概率密度 probability density
概率密度函数 probability density function
概率曲线 probability curve
均匀分布 uniform distribution
指数分布 exponential distribution
指数分布密度函数 exponential distribution density function
正态分布,高斯分布 normal distribution
标准正态分布 standard normal distribution
正态概率密度函数 normal probability density function
正态概率曲线 normal probabi-lity curve
标准正态曲线 standard normal curve
柯西分布 Cauchy distribution
分布密度 density of distribution

2.5.3 多维随机变量及其分布 (Multivariate Random Variables and Distributions)

二维随机变量 two-dimensional random variable
联合分布函数 joint distribution function
二维离散型随机变量 two-dimensional discrete random variable
二维连续型随机变量 two-dimensional continuous random variable
联合概率密度 joint probability variable
n 维随机变量 n-dimensional random variable
n 维分布函数 n-dimensional distribution function
n 维概率分布 n-dimensional probability distribution
边缘分布 marginal distribution
边缘分布函数 marginal distribution function
边缘分布律 law of marginal distribution
边缘概率密度 marginal probability density
二维正态分布 two-dimensional normal distribution
二维正态概率密度 two-dimensional normal probability density
二维正态概率曲线 two-dimensional normal probability curve
条件分布 conditional distribution
条件分布律 law of conditional distribution
条件概率分布 conditional probability distribution
条件概率密度 conditional probability density
边缘密度 marginal density
独立随机变量 independent random variables

2.5.4 随机变量的数字特征 (Numerical Characteristics of Random Variables)

数学期望,均值 mathematical expectation
期望值 expectation value
方差 variance

标准差 standard deviation

随机变量的方差 variance of random variables

均方差 mean square deviation

相关关系 dependence relation

相关系数 correlation coefficient

协方差 covariance

协方差矩阵 covariance matrix

切比雪夫不等式 Chebyshev inequality

2.5.5 大数定律及中心极限定理（Law of Large Numbers and Central Limit Theorem）

大数定律 law of large numbers

切比雪夫定理的特殊形式 special form of Chebyshev theorem

依概率收敛 convergence in probability

伯努利大数定律 Bernoulli law of large numbers

同分布 same distribution

列维－林德伯格定理、独立同分布中心极限定理 independent Levy-Lindberg theorem

辛钦大数定律 Khinchine law of large numbers

李亚普诺夫定理 Liapunov theorem

棣莫弗－拉普拉斯定理 De Moivre-Laplace theorem

2.5.6 样本及抽样分布（Samples and Sampling Distributions）

统计量 statistic

总体 population

个体 individual

样本 sample

容量 capacity

统计分析 statistical analysis

统计分布 statistical distribution

统计总体 statistical ensemble

随机抽样 stochastic sampling/random sampling

随机样本 random sample

简单随机抽样 simple random sampling

简单随机样本 simple random sample

经验分布函数 empirical distribution function

样本均值 sample average/sample mean

样本方差 sample variance

样本标准差 sample standard deviation

标准误差 standard error

样本 k 阶矩 sample moment of order k

样本中心矩 sample central moment

样本值 sample value

样本大小、样本容量 sample size

样本统计量 sampling statistics

随机抽样分布 random sampling distribution

抽样分布、样本分布 sampling distribution

自由度 degree of freedom

Z 分布 Z-distribution

U 分布 U-distribution

2.5.7 参数估计（Parameter Estimations）

统计推断 statistical inference

参数估计 parameter estimation

分布参数 parameter of distribution

参数统计推断 parametric statistical inference

点估计 point estimate/point estimation

总体中心距 population central moment

总体相关系数 population correlation

coefficient
 总体分布 population distribution
 总体协方差 population covariance
 点估计量 point estimator
 估计量 estimator
 无偏估计 unbiased estimate/unbiased estimation
 估计量的有效性 efficiency of estimator
 矩法估计 moment estimation
 总体均值 population mean
 总体矩 population moment
 总体 k 阶矩 population moment of order k
 总体参数 population parameter
 极大似然估计 maximum likelihood estimation
 极大似然估计量 maximum likelihood estimator
 极大似然法 maximum likelihood method/maximum-likelihood method
 似然方程 likelihood equation
 似然函数 likelihood function
 区间估计 interval estimation
 置信区间 confidence interval
 置信水平 confidence level
 置信系数 confidence coefficient
 单侧置信区间 one-sided confidence interval
 置信上限 confidence upper limit
 置信下限 confidence lower limit
 U 估计 U-estimator
 正态总体 normal population
 总体方差的估计 estimation of population variance
 置信度 degree of confidence
 方差比 variance ratio

2.5.8 假设检验（Hypothesis Testing）

 假设的 hypothetical
 先验的 priori
 后验的 posteriori
 参数假设 parametric hypothesis
 假设检验 hypothesis testing
 两类错误 two types of errors
 统计假设 statistical hypothesis
 统计假设检验 statistical hypothesis testing
 检验统计量 test statistics
 显著性检验 test of significance
 统计显著性 statistical significance
 单边检验、单侧检验 one-sided test
 单侧假设、单边假设 one-sided hypothesis
 双侧假设 two-sided hypothesis
 双侧检验 two-sided test
 显著水平 significant level
 拒绝域/否定区域 rejection region
 接受区域 acceptance region
 U 检验 U-test
 F 检验 F-test
 方差齐性的检验 homogeneity test for variances
 拟合优度检验 test of goodness of fit

3 力　学

3.1 运　动

3.1.1 基本概念

维度 dimension
质点 particle
刚体 rigid body
原点 origin
推动 impetus
位移 displacement
速度 velocity
加速度 acceleration
位置向量 position vector
牛顿第一运动定律 Newton's first law
力 force
质量 mass
惯性系统 inertial frame
牛顿第二运动定律 Newton's second law
平衡 equilibrium
牛顿第三运动定律 Newton's third law
张力 tension
摩擦力 friction force
内摩擦 internal friction
离心力 centripetal force
重力 gravity
点力 point force
线力 line force
面力 plane force
体力 body force
浮力 buoyancy
阻塞力 blocked force
集中力 concentrated force
惯性力 inertial force
测力计 force gauge
动量 momentum
动量守恒 conversation of momentum
功 work
能量 energy
功能原理 work-energy theorem
势能 potential energy
振荡 oscillation
简谐运动 simple harmonic motion
胡克定律 Hooke's Law
劲度系数 spring constant
弹回 resilience
弹力 elasticity
振幅 amplitude/magnitude
周期 period
频率 frequency
角频率 angular frequency
力矩 moment of force
扭矩 torsional moment
惯性矩 moment of inertia
转动惯量 rotational inertia
角加速度 angular acceleration
顶 crest
低谷 trough
顶点 vertex
运动学上地 kinematically
随时间的变化 time history
按时间平均的 time-averaged
时间相依 time-dependent
时谐波 time-harmonic wave

3.2 应力、应变与材料特性

材料密度 material density
物料性质 material properties

各向同性 isotropic
各向异性 anisotropic
安全系数 safety factor
应力 stress
 正应力 normal stress
 剪切力 shear force
 剪应力 shear stress
 主应力 principal stress
 屈服应力 yield stress
应变 strain
弹性模量 elastic modulus
泊松比 Poisson's ratio
塑性变形区 plastically deforming area
预应力的(混凝土) prestressed
剪切模量 shear modulus
剪切速度 shear velocity
横切力 transverse shear force

3.3 固体力学

干摩擦 dry friction
库伦摩擦 Coulomb friction
滚动 rolling motion
滑动 sliding motion
框架 framework
自由体 free body
刚体 rigid body
受力图 free-body diagram
悬臂 cantilever
固支 clamped
简支 simple support
铰支 hinged
滑动 slide
斜面 slope
弯曲变形 bending deflection
弯曲弹性 bending elasticity
弯曲模量 bending modulus
弯曲刚度 bending stiffness

弯矩 bending moment
控制力矩 control moment
扭力常数 torsional constant
扭力的 torsional
复合杆 coupled pole
耦合系统 coupled system
弹塑性 elasto
弹性动力学 elastodynamics
人造橡胶 elastomer
人造橡胶的 elastomeric
抗挠刚度 flexural rigidity
岩土工程 geotechnical engineering
简支梁 simply supported beam
Ⅰ型梁 I-beam
非弹性屈曲 inelastic buckling
易弯的 limp
瑞利 Rayleigh
静变位,静荷载挠度 static deflection

3.4 振 动

振动的 vibrant
振动的 vibratory
自由振动 free vibration
平衡位置 equilibrium position
声振环境 vibroacoustic environment
波源 wave source
波腹 waveloop, antinode
波节 wave node
机械波 mechanical wave
波前,波阵面 wavefront
波导 waveguide
波数滤波器 wavenumber filter
衰减振动 attenuation vibration
共振 resonance
瞬态解 transient solution
稳态解 steady state solution
位移共振频率 displacement

3 力 学

resonance frequency

力阻抗 force impedance

力阻 force resistance

力抗 force resistance

质量抗 mass resistance

弹性抗 elastic resistance

速度共振频率 speed resonance frequency

加速度共振频率 acceleration resonance frequency

质量控制区 mass control area

弹性控制区 elastic control area

力阻控制区 force control area

振动计 vibrometer

隔振 vibration isolation

拾振 vibration pickup

简正频率 simple frequency

基频 fundamental frequency

泛频 overtones

谐频 harmonic frequency

谐频发生 harmonic frequency generation

驻波 standwave

反共振 anti-resonance

零阶柱贝塞尔方程 zero-order Bessel equation

4 声 学

4.1 基本概念

声学 acoustic
空气声学 aeroacoustics
生物声学 bioacoustics
流体动力学 fluid dynamics
空气动力学的 aerodynamic
声 sound
声全息 sound holography
声压 sound pressure
瞬时声压 instantaneous sound pressure
峰值声压 peak sound pressure
有效声压 effective sound pressure
纵波 longitudinal waves
横波 transversal waves
横向的 transverse
横着地 transversely
干涉 interference
反射 reflection
散射 scatter
衍射 diffraction
折射 refraction
入射法线 incident normal
入射波 incident wave
正入射 normal incidence
掠入射 grazing incidence
平面外 out-of-plane
反射波 reflected wave
驻波 stand wave
反相 antiphase
破坏性干涉 destructive interference
建设性干涉 constructive interference

声共振 acoustic resonance
透射截面 transmission cross section
发射束 transmitted beam
线性波动方程 linearised wave equation
声速 speed of sound
绝对温度 absolute temperature
绝热压缩 adiabatic compression
均匀压缩 uniform compression
膨胀 inflation
绝热的 adiabatic
等温的 isothermal
边界条件 boundary condition
欧拉运动方程 Euler's equation of motion
平面波 plane wave
简谐波 harmonic wave
泛音 overtones
角频率 angular frequency
波数 wavenumber
波长 wavelength
声场 sound field
  自由场 free field
  近场 near field
  远场 far field
粒子速度 particle velocity
反射因子 reflection factor
相移 phase shift
驻波场 standing wave field
相干波 coherent wave
驻波比 standing wave ratio
音调 pitch
节拍 beats
音色 timbre
纯音 pure-tone
响度级 loudness level
球形声波 spherical sound waves
反距离法则 the inverse distance law
正交 quadrature
反应场 reactive field

4.2 声测量

声压 sound pressure
超声波 ultrasonic wave
次声波 infrasonic wave
远超过音速的 hypersonic
八度 octave
随机噪声 random noise
白噪声 white noise
粉红噪声 pink noise
分贝 decibel
声压级 sound pressure level
声强 sound intensity
声强级 sound intensity level
声功率 sound power
声能量 sound energy
声功率级 sound power level
声级计 sound level meter
电容式麦克风 condenser microphones
自由场校正 free field correction
全向 omnidirectional
A 声级 A-weighted sound pressure level
间歇性噪声 intermittent noise
声曝级 sound exposure level
热通量密度 heat flux density
声势能 acoustic potential energy
声模态 acoustic mode
测功计 dynamometer
倍频带 octave band
互易律 reciprocity law
强度放大 intensity amplification
声功率吸收系数 sound power absorption coefficient
声功率计算 sound power calculation
声功率值 sound power value
声能量密度 sound energy density
声强法 sound intensity method
自由场法 free-field method
消声室 anechoic room
混响室 reverberation room
室内混响 indoor reverberation
混响时间 reverberation time
混响半径 reverberation radius
直达声场 direct sound field
无回声的 anechoic
声吸收 sound absorption
吸收系数 absorption coefficient
瞬态互调失真 transient intermodulation distortion

4.3 阻抗

阻抗 impedance
导纳 admittance
机械阻抗 mechanical impedance
声阻抗 acoustic impedance
体积速度 volume velocity
测定体积的 volumetric
辐射阻抗 radiation impedance
亥姆霍兹共振器 helmholtz resonator

4.4 流体动力学

流体 fluid
流体的 fluidic
理想流体 ideal fluid
拟理想 quasi-ideal
准静态 quasi-static
塑性 plastic
伪塑性 pesudo-plastic
宾汉塑性流体 Bingham plastic fluid
缓释物质 dilatant substances
触变性物质 thixotropic substances
流凝性物质 rheopectic substances

黏弹性材料 viscoelastic materials
黏滞阻力 viscous resistance
计算流体力学 computational fluid dynamic(CFD)
对流的 convective
液压的 hydraulic
流量模数 hydromodulus
比热 specific heat
水汽 vapor
静止 quiescent
质量守恒 mass conservation
动量守恒 momentum conservation
能量守恒 energy conservation
熵 entropy
等熵 isentropy
同质 homogeneous
正压 barotropic
扩张率 dilatation rate
可压缩流体 compressible fluid
不可压缩流体 incompressible fluid
无旋流 irrotational flow
斯特劳哈尔数 strouhal number
动黏滞率,动黏度 kinematic viscosity
热扩散率 heat diffusivity
雷诺数 Reynolds number
等温 isothermal
热对流 thermal convective current
不均匀性 inhomogeneity
焓 enthalpy
伯努利常数 bernoulli constant
涡度 vorticity
涡旋声理论 vortex sound theory
黏的 viscous

4.5 声辐射

点源 point sources
单极子 monopole
偶极子 dipole
四极子 quadrupole
多极子 multipole
脉动声源 pulsating sound source
互易原理 reciprocity principle
远场近似 far-field approximation
瑞利积分 Rayleigh's integral
方向性因子 directivity factor
多普勒效应 Doppler effect
接受率 receptance rate
封闭空间 enclosure space
围场 enclosure filed
扩散声场 diffuse acoustic filed
辐射模态 radiation mode
自辐射阻抗 self-radiation impedance
声波的透射 transmission of sound waves
声压的反射系数与透射系数 reflection/transmission coefficient of sound pressure
硬边界 hard boundary
软边界 soft boundary
斯奈尔声波反射与折射定律 Snell acoustic reflection and refraction law
全透射 full transmission
全反射 total reflection(TIR)
隔声量 sound insulation
隔声量测试 sound insulation test
声透射系数,传声系数 acoustic transmission coefficient
吸声系数 sound absorption coefficient
输入阻抗 input impedance
辐射阻抗 radiation impedance
自辐射阻抗 self-radiation impedance
互辐射阻抗 mutual radiation impedance
辐射的指向特性 directional characteristics of radiation

5 自动控制

5.1 基本概念

5.1.1 发展历史

经典控制理论 classic control theory
　　时域法 time domain method
　　复域法 complex domain method
　　频域法 frequency domain method
现代控制理论 modern control theory
　　线性系统 linear system
　　最优控制 optimal control
　　最佳估计 best estimate
　　系统辨识 system identification
　　自适应控制 adaptive control
　　鲁棒控制 robust control
　　容错控制 fault tolerant control
　　分布式控制 distributed control
　　大系统复杂系统控制 large-scale complex system control
智能控制理论 intelligent control theory
　　专家系统 expert control
　　模糊控制 fuzzy control
　　神经网络 neural network control
　　遗传算法 genetic algorithm
　　蚁群算法 ant colony algorithm
　　模拟退火 simulated annealing

5.1.2 组成

自动控制 automatic control
自动控制系统 automatic control system
系统 system
控制信号 control signal, manipulated variable
干扰 disturbance
给定元件 designate component
比较元件 compare component
放大元件 amplify component
执行元件 execute component
　　电动作动器 electric actuator
　　电磁作动器 electromagnetic actuator
　　气动作动器 pneumatic actuator
　　液压作动器 hydraulic actuator
测量元件 measure component
校正元件 correction point component
开环控制 open-loop control
闭环控制 closed-loop control

5.1.3 控制类型

单输入单输出 single-input/single-output(SISO)
多输入多输出 multi-input/multi-output(MIMO)
线性系统 linear system
　　线性时不变系统 linear time invariant system
　　线性时变系统 linear time variant system
　　叠加原理 superposition principle
　　齐次原理 homogeneous principle
非线性系统 nonlinear system
　　继电特性 relay characteristics
　　死区特性 dead zone characteristics
　　饱和特新 saturation characteristic
　　间隙特性 gap characteristics
　　摩擦特性 friction characteristics
本质非线性 intrinsic nonlinearity
开环控制 open-loop control
连续系统 continuous system

离散系统 discrete system
连续信号 continuous signal
离散信号 discrete signal
恒值系统 constant-value system
伺服系统 servo system
  逻辑控制 logic control
  顺序控制 sequence control
闭环控制 close-loop control
前馈控制 feedforward control
反馈控制 feedback control
单通道控制 single channel control
多通道控制 multi channel control
实时控制 real-time control
在线控制 on-line control
离线控制 off-line control

5.1.4 性能指标

控制目标 control object
  稳定性 stability
  准确性 accuracy
  快速性 rapidity
静态 static state
稳态 stable state
暂态 transient state
动态 dynamic state
延时 delay time
上升时间 rise time
峰值时间 peak time
最大超调 maximum overshoot
调整时间 settling time
振荡次数 number of oscillations
稳态响应 steady-state response
绝对稳定 absolute stability
相对稳定 relative stability
稳态误差 steady-state error

5.2 模 型

数学模型 mathematical model
静态模型 static model
动态模型 dynamic model
参数模型 parametric model
非参数模型 nonparametric model
白箱模型 white box model
灰箱模型 gray box model
黑箱模型 black box model
微分方程 differential equation
  常微分方程 ordinary differential equation
  偏微分方程 partial differential equation
差分方程 difference equation
瞬态响应函数 transient response function
传递函数 transfer function
频率响应函数 frequency response function
非线性系统的线性化 linearization of nonlinear system
  泰勒展开 Taylor expansion
  稳定工作点 stable operating point
典型输入信号 classic input signal
  阶跃函数 step function
  斜坡函数 slope function
  脉冲函数 impulse function
  抛物线函数 parabolic function
  正弦函数 sine function
拉普拉斯变换 Laplace transform
拉普拉斯反变换 Laplace inverse transform
积分变换 integral transform
傅里叶变换 Fourier transform
留数法 residual method

零点 zero point
极点 pole point
卷积 convolution
方块图 block diagram
信号流图 signal flow diagram
开环传递函数 open-loop transfer function
闭环传递函数 closed-loop transfer function
正向通道 forward channel
反馈通道 feedback channel
梅森公式 Meson function

5.3 系 统 分 析

5.3.1 时域分析

定性的 qualitative
定量法 quantifying
暂态响应分析 transient response analysis
时域分析 time-domain analysis
脉冲响应函数 impulse response function
过阻尼系统 over damp system
临界阻尼系统 critical damp system
欠阻尼系统 under damp system
无阻尼系统 undamped system
稳定性分析 stability analysis
　　稳定 stable
　　相对稳定 relative stable
　　临界稳定 critical stable
　　不稳定 unstable
劳斯稳定判据 Routh's stability criterion
赫尔维茨稳定判据 Hurwitz's stability criterion
特征方程 characteristic equation

误差传递函数 error transfer function
静态位置误差系数 static position error constant
静态速度误差系数 static velocity error constant
静态加速度误差系数 static acceleration error constant

5.3.2 根轨迹法

根轨迹法 root locus method
闭环系统的主导极点 dominant closed-loop poles
广义根轨迹 generalized root locus
分离点 separation point
会合点 meeting point
渐近线 asymptotic line

5.3.3 频率特性分析

极坐标 polar coordinates
最小相系统 minimum-phase system
频率响应方法 frequency-response method
伯德图 Bode diagram
对数图 logarithmic plot
谐振频率 resonant frequency
谐振峰值 resonant peak value
奈奎斯特稳定准则 Nyquist stability criterion
奈奎斯特曲线 Nyquist plot
映射理论 mapping theorem
相角裕度 phase margin
幅值裕度 gain margin
截止频率 cutoff frequency
带宽 bandwidth
反馈补偿 feedback compensation
零点配置方法 zero-placement approach
非最小相系统 nonminimum-phase

system
  超前补偿 lead compensator
  滞后补偿 lag compensator
  滞后-超前补偿 lag-lead compensation
  降低稳定性 de-stabilize
  相位复(数)矢量 phasor

5.3.4 非线性系统

  相平面法 phase plane method
  描述函数法 description function method
  相轨迹 phase trajectory
  奇点 singularity/singular point
  奇线 odd line
  极限环 limit cycle

5.4 离散系统

  离散系统 discrete-time system
  采样控制系统 sampling control system
  数字控制系统 digital control system
  A/D 转换器 A/D converter
  D/A 转换器 D/A converter
  香农采样定理 Shannon's sampling theorem
  Z 变换 Z-transformation
  Z 反变换 Z inverse transformation
  朱利判据 Julie criterion

5.5 现代控制理论

  状态 state
  状态变量 state variables
  状态向量 state vector
  状态空间 state space
  状态空间表达式 state-space equation
  串联分解 series decomposition
  并联分解 parallel decomposition
  可控标准型 controllable canonical form
  可观标准型 observable canonical form
  约当标准型 Jordan canonical form
  齐次方程的求解 solution of homogeneous state equation
  零输入解,零输入响应 zero input response
  零状态响应 zero state response
  矩阵指数 matrix exponential
  状态转移矩阵 state-transition matrix
  状态转移矩阵的性质 properties of state-transition matrix
  非齐次方程的求解 solution of nonhomogeneous state equation
  最优控制 optimal control
  性能函数 performance function
  凯莱-哈密顿定理 Cayley-Hamilton theorem
  最小多项式 minimal polynomial
  能控性 controllability
  状态能控 state controllability
  输出能控 output controllability
  能观性 observability
  能控性分解 controllability decomposition
  能观测性分解 observable decomposition
  对偶性 principle of duality
  自治系统 autonomous system(AS)
  平衡状态 equilibrium state
  一致稳定 consistent stability
  渐近稳定 asymptotically stable
  大范围渐近稳定 Large-scale asymptotic stability
  李亚普诺夫第一方法 Lyapunov's first approach
  李亚普诺夫第二方法 Lyapunov's second approach
  状态反馈 state feedback

5 自动控制

反馈增益矩阵 feedback gain matrix
状态观测器 state observer
双位置、开关控制 two-position/on-off control
非线性控制 nonlinear control
定性控制 qualitative control
预测控制 predictive control

6 电力电子

6.1 基本概念

电路 circuit
电荷 charge
电流 current
    直流电 direct current(DC)
    交流电 alternating current(AC)
电位 electric potential
电压 voltage
电势 potential
电阻 resistance
电感 inductance
    自感 self-inductance
    互感 mutual inductance
感应的 induced
电容 capacitance
电源 power
    直流电源 DC power supply
    交流电源 AC power supply
电位计 potentiometer
验电板 proof-plane
负载 load
电极 electrode
静电的 electrostatic
超导现象 superconducting phenomenon
装有保险丝的 fused
电功率 electric power
电功率计 electro-dynamometer

6.2 电路计算

欧姆定律 Ohm's law
电压源 voltage source
电流源 current source
受控源 controlled source
基尔霍夫定律 Kirchhoff's law
支路 branch
节点 node
回路 loop
参考节点 reference node
非关联方向 non-affiliated direction
网孔 mesh
串联 series connection
并联 parallel connection
伏安曲线 voltage-current curve
输入电阻 input resistance
支路电流法 branch current method
网孔电流法 mesh current method
网状的 reticulated
回路电流法 loop current method
叠加原理 superposition principle
储能元件 energy storage components
阻抗 impedance
导纳 admittance
复阻抗 complex impedance
电抗 reactance
复导纳 complex admittance
电纳 susceptance
感性 inductive
容性 capacitive
瞬时功率 instantaneous power
平均功率 averaging power

6.3 电　子

电子管 eletrocnic tube
晶体管 transistor
集成电路 integrated circuit
导体 conductor
绝缘体 insulator
semiconductor
本征半导体 intrinsic semiconductor
电子-空穴对 electron hole pair
杂质半导体 impurity semiconductors
半导体二极管 diode
PN 结 PN junction
半导体三极管 semiconductor transistor
特性曲线 characteristic curve
场效应管 field effect transistor
放大器 amplifier
放大电路 amplification circuit
静态工作点 static operating point
截止失真 cut off distortion
非线性失真 nonlinear distortion
饱和失真 saturation distortion
零点漂移问题 zero drift problem
集成运算放大器 integrated operational amplifier
半导体集成电路 semiconductor integrated circuit
混合集成电路 hybrid integrated circuit
双极型集成电路 bipolar integrated circuit
模拟集成电路 analog integrated circuit
数字集成电路 digital integrated circuit
偏置电路 bias circuit
差模信号 differential mode signal
共模信号 common-mode signal
共模抑制比 common-mode rejection ratio(CMRR)
正弦波振荡器 sine wave oscillator
RC 桥式(文氏)振荡器 RC bridge type (venturi) oscillator
整流 rectification
半波整流 half wave rectification
全波整流 full-wave rectification
整流器 rectifier
桥式整流器 bridge rectifier
倍压整流 doubling rectifier
稳压 regulator
数字信号 digital signal
数字电路 digital circuit
门电路 gate circuit
组合电路 combination circuit
时序电路 sequential circuit
ASCII 代码 American Standard Code for Information Interchange
二进制补码 two's complement
逻辑变量 logical variable
逻辑函数 logical function
卡诺图 Karnaugh map
编码器 coder, encoder
译码器 decoder
触发器 flip-flop
同步 RS 触发器 synchronous RS flip-flop
主从 RS 触发器 master-slave RS flip-flop
主从 JK 触发器 master-slave JK flip-flop
T 触发器 T flip-flop
移位寄存器 shift register
字节 byte
计数器 counter
脉冲电路 pulse circuit

限幅电路 limiting circuit
多谐振荡器 multivibrator
正弦信号 sinusoidal signal
零输入响应 zero input response
起始状态 start state
零状态响应 zero state response
傅里叶变换 Fourier transform
拉普拉斯变换 Laplace transform
初值定理 initial value theorem

6.4 运算电路

反馈放大器 feedback amplifier
深度负反馈 depth negative feedback
比例器 proportional controller
减法器 subtractor
积分器 integrator
半加器 half-adder
全加器 full-adder
数值比较器 numerical comparator
信号发生器 signal source
集中参数电路 lumped-parameter circuit
分布参数电路 distributed constant (parameter) circuit
全补偿运算放大器 fully-compensated operational amplifier
隔离放大器 isolation amplifier
（无线电）前置放大器 preamplifier
振荡电场 oscillating electric field
移相器 phase shifter
相位滞后网络 phase-lag network

6.5 电　力

功率因数 power factor
无功功率 reactive power
视在功率 apparent power
有功功率 active power
复功率 complex power
幅频特性 amplitude-frequency characteristics
相频特性 phase-frequency characteristics
串联谐振 series resonance
并联谐振 parallel resonance
三相电路 three-phase A. C. circuit
对称三相电源 symmetrical three-phase power supply
三相四线制 three phase four wire system
线电压 line-to-line voltage
相电压 phase-to-neutral voltage
线电流 wire current
相电流 phase current
未绝缘的 uninsulated

7 传感器和作动器

7.1 基本概念

灵敏元件 sensitive element
传感器 sensor, transducer
传感器考量 sensor considerations
  测量本质 nature of measure
    辐射 radiation
    磁性 magnetic
    热 thermal
    机械 mechanical
    化学 chemical
  传感器输出 sensor output
    热 thermal
    磁性 magnetic
    电 electric
    光 optical
    机械 mechanical
  环境 environment
    腐蚀 corrosive
    热 thermal
    磁 magnetic
    电 electric
  界面 interface
    尺寸 size
    几何 geometry
    机械性能 mechanical properties
  操作性能 operational properties
    灵敏度 sensitivity
    带宽 bandwidth
    线性度 linearity
    量程 operational range
    标距长度 gauge length
作动器 actuator
作动器考量 actuator considerations
  驱动能量的本质 nature of driving energy
    光学 optical
    磁性 magnetic
    热 thermal
    机械 mechanical
    化学 chemical
    电 electric
  作动器特性 properties of the actuator
    位移 displacement
    力生成 force generation
    磁滞 hysteresis
    响应时间 response time
    带宽 bandwidth
  激励信号 stimulus
    应力 stress
    应变 strain
    压力 pressure
    温度 temperature
    光 light
    气体分子 gas molecules
    电场 electrical field
    磁场 magnetic field
    固有的 intrinsic
    外在的 extrinsic

7.2 传 感 器

7.2.1 基本概念

传感器类型 types of sensors
  模拟 analog
  数字 digital
  主动传感器 active sensor

被动 passive sensor
传感器输出形式 the form of sensor output
  阻抗变化 resistance change
  电压变化 voltage change
  电容变化 capacitance change
  电流变化 current change
传感器品质 quality of a sensor
  分辨率 resolution
  准确度 accuracy
  精度 precision
  重复性 repeatability
  量程 range
  跨度 span
  稳定性 stability
  死区 dead zone
  间隙 backlash
  磁滞 hysteresis
误差 error
  绝对误差 absolute error
  相对误差 relative error
传感器特性 sensor characteristics
  静态特性 static characteristics
  动态特性 dynamic characteristics
  响应时间 response time
  时间常数 time constant
  上升时间 rise time
  调整时间 settling time
常见的可以测量的现象 commonly detectable phenomena
  温度 temperature
  机械运动 mechanical motion
  光 optical
  电 electrical
  化学 chemical
  电磁 electromagnetic
  生物 biological
  放射性 radioactivity

所使用的物理原理的例子 examples of physical principles used
  安培定律 Amperes's law
  居里·韦斯定律 Curie-Weiss law
  法拉第诱导定律 Faraday's law of induction
  光导效应 photoconductive effect
惯性传感器 inertial sensor

7.2.2 分类

位移传感器 displacement sensor
  阻抗 resistive
    电位计 potentiometers
    电感 inductance
    线性可变差动变压器 linear variable differential transformer
  临近传感器 proximity sensors
    非接触 non-contact
    编码器 encoders
    电容传感器 capacitive sensors
应变计 strain sensor
  硅应变计 silicon strain gauges
  压电应变计 piezoelectric strain gauges
  光纤应变计 fibre-optic strain gauges
  弯曲应变 bending strain
  剪切应变 shear strain
  扭转应变 torsional strain
  泊松应变 Poisson strain
力传感器 force sensor
加速度传感器 acceleration sensor
  压电 piezoelectrics
温度传感器 temperature sensor
  双金属条 bimetallic strips
  流体膨胀装置 fluid expansion devices
  状态变化温度测量 change-of-state temperature measurement
  热电偶 thermocouple

电阻温度器件 resistance temperature devices(RTD)

红外设备 infrared devices

功率传感器 power sensing device

7.2.3 智能传感器系统

智能材料 smart materials

智能系统 smart system

补偿 compensation

自诊断 self-diagnostics

自校准 self-calibration

自适应 self-adaptation

计算 computation

信号调理 signal conditioning

数据缩减 data reduction

触发事件的检测 detection of trigger events

通信 communications

网络协议标准化 network protocol standardization

集成 intergration

传感和计算在芯片级上耦合 integration coupling of sensing and computation at the chip level

微机电系统 micro electromechanical systems(mems)

其他 others

多模式 multi-modal

多维 multi-dimensional

多层 multi-layer

主动 active

自感知 autonomous sensing

7.3 作 动 器

7.3.1 基本概念

作动 actuation

作动器材料 actuator materials

形状记忆合金 shape memory alloys

磁致伸缩材料 magnetostrictive materials

压电和电致伸缩材料 piezoelectric & electrostrictive materials

电流变液和磁流变液 electrorheological & magnetorheological fluids

显示形状记忆效应的合金系列 alloy families that exhibit shape memory effect

铜-铝-镍 copper-aluminum-nickel

铜-锌-铝 copper-zinc-aluminum

铁-锰-硅 iron-manganese-silicon

镍钛(镍钛诺) nickel-titanium (nitinol)

奥氏体 austenite

马氏体 martensite

宾汉塑性体 Bingham body

宾汉模型 Bingham model

稀土超磁致伸缩材料 terfenol-D

各向异性的 anisotropic

流变学的 rheological

机电的 electromechanical

音圈 voice coil

共振式消声器 resonant muffler

7.3.2 其他作动形式

现成的 off-the-shelf

线圈阻尼 coil resistance

抗磁力 coercive force

磁单极子 monopole

剩磁 remanence

气动系统 pneumatic system

由压缩空气操作(推动)的 pneumatic

由空气作用 pneumatically
压缩 compression
空气弹性变形的 aeroelastic

7.4 压电材料

压电效应 piezoelectric effect
    直接压电效应 direct piezoelectric effect
    逆压电效应 converse piezoelectric effect

类型(type)	材料(materials)
单晶体 single crystals	铅铌酸盐 lead magnesium niobate(PMN)
陶瓷 ceramics	锆钛酸铅 lead zirconate titanate(PZT)
	铌酸铅 lead meta niobate(LMN)
	钛酸铅 lead titanate(LT)
	铅铌酸盐 lead magnesium Niobate(PMN)
聚合物 polymers	聚偏二氟乙烯 polyvinylidene difluoride(PVDF)
复合材料 composites	陶瓷聚合物 ceramic-polymer
	陶瓷玻璃 ceramic-glass

压电陶瓷 piezoceramic
正向的 positive-going
锆钛酸铅 lead zirconate titanate(PZT)

7.5 其他材料

银 silver
镍 nickel
铽(符号 Tb) terbium
钛 titanium
蜡 wax
铅甲酸盐 leadmetaniobate(LMN)
铌酸钠 sodium potassium niobate(SPN)
氯丁(二烯)橡胶 neoprene
聚氨酯 urethane
镍钛诺(镍和钛的合金) nitinol
塑胶 perspex
聚酰胺 polyamide
聚亚安酯 polyurethane
聚偏二乙烯的 polyvinylidene
(聚)硅酮 silicone
铁酸盐 ferrite
高分子材料 polymers
聚类 clustering
功能材料 functional materials
玻璃纤维 fiberglass
氟化物 fluoride
居里温度 Curie temperature
连续的改变 flux
黏塑(性)流 viscoplastic flow
成分 constituent
电介质 dielectric
隔膜 diaphragm
中性化器 neutralizer
双压电晶片元件 bimorph
介电常数,电容率 permittivity
多边化,多角形化 polygonization
抗扯强度 tear strength
序列,排列 permutation
薄片 wafer
细丝 filament
薄膜 film
泡沫 foam
蜂窝 honeycomb

7 传感器和作动器

薄片 laminae
小片 patch
（夹板的）层片 ply
伸展 extrusion
延伸 extension
（将薄片砌合在一起）制成（材料）laminated
不齐 irregularities
压缩态 squeezed state
堆 stack
极化 polarization
极化区域 polarized area
交织 interleave
改变…的形状 reshape
把…夹在…之间 sandwich
覆盖 overlaid
在…上铺或盖 overlay
使（某物）被遮暗 overshadow
可压碎的 crushable
晶体的 crystallographic
内长的 endogenous
外生的 exogenous
外部的 extraneous
铁磁的 ferromagnetic
石墨的 graphite
酚的,石碳酸的 phenolic
可渗透的 permeability
同性质的 homogeneous
极微小的 infinitesimal
中间的 intermediate
不对称的 irregular
等方性的 isotropic
肉眼可见的 macroscopically
半透明的 translucent
不透明的 opaque
能穿透的 porous
易燃的 flammable
剪裁讲究的 tailored

粒子数增加的 populated
顶盖的,覆膜的 tectorial
由薄片或层状体组成的 laminar

8 仿真计算

8.1 数字信号处理

信息 information
信号 signal
确定性信号 deterministic signal
随机信号 stochastic signal
连续(时间)信号 continuous(time) signal
标量量化 scalar quantization
矢量量化 vector quantization
编码 coding
动态系统 dynamic system
因果系统 causal system
非因果系统 noncausal system
宽带 broad band
遍历性的 ergodic
各态历经 ergodicity
范围,光谱 spectra
谱密度 spectral density
谱因子分解 spectral factorization
谱性质 spectral properties
范围 spectrum
使平滑 smoothen
单位方差 unit variance
零相移 zero phase shift
零交点 zero-crossing
过筛 sifting
调制 modulation
巴特沃斯滤波器 Butterworth filter
自相关 auto-correlation
互相关 cross-correlation
自回归 autoregressive

分块系统 block Toeplitz system
类比插值法 analogue interpolation
渐近有效估计 asymptotically efficient estimate
倒谱分析 cepstral analysis
倒谱距离 cepstral distance
一致 coherence
条件数 condition number
共形阵 conformal array
互谱密度 cross spectral density
周期的 cyclic
失真 distortion
截断的 truncate
切断 truncation
舍入误差 rounding error
空间混叠 spatial aliasing
随机的 stochastic

8.2 控制算法

模振幅 mode amplitude
振形 mode shape
模空间 moduli space
系数 modulus
多尺度分析 multiscale analysis
多级控制系统,分级控制系统 multi-level control system
单调阻尼稳定点 monotonically damped stable point
模内流动性 mold flow
预定的极限 pre-established limit
平方律误差 square law error
二乘法 square law
奇异值分解 singular value decomposition (SVD)
同步系统 simultaneous system
边(频)带 sideband
递归最小二乘方 RLS recursive

least square
　　重复(迭代)计算 repetitive computation
　　拟弹性近似 pseudo-elastic approximation
　　伪逆 pseudo-inverse
　　参数化 parametrization
　　模型搜索 pattern search
　　拓扑 topology
　　感知器(模拟人类视神经控制系统的图形识别机) perceptron
　　参数辨识 parameter identification
　　重叠保留法 overlap-save method
　　原系统 original system
　　节线 nodal line
　　节的 nodal
　　修改 modify
　　模块性 modularity
　　计算上不可能的 computationally infeasible
　　约束 constrain
　　流程图 flow chart
　　预期的 contemplated
　　控制系统 control system
　　因果关系 causa
　　汉克尔函数 hankel function
　　厄米二次型 Hermitian quadratic form
　　厄米共轭 Hermitian conjugation
　　海赛矩阵 Hessian matrix
　　惯性矩阵 inertial matrix
　　托普利茨矩阵 Toeplitz matrix
　　线性回归 linear regression
　　极小值 minimum, minima(minimum 的复数)
　　极小化极大算法 minimax
　　可传送性 transmissibility
　　边界刚度矩阵 boundary stiffness matrix
　　级联的 cascaded

　　特征长度 characteristic length
　　特征定义 characterizing definition
　　混沌的 chaotic
　　同位置配置法 collocation method
　　同位置的 collocated
　　转角频率 corner frequency
　　截止角频率 cut-off angular frequency
　　截止频率 cut-on frequency
　　迅速消失遗忘的 evanescent
　　发展 evolve
　　穷举搜索 exhaustive search
　　爆炸的 explosive
　　指数式稳定 exponentially stable
　　推算 extrapolate
　　极值 extrema(extremum 的复数)
　　局部的 local
　　全局的 global
　　频率窗口 frequency bin
　　杂交生成的生物体 hybrid
　　内模控制 internal modal control(IMC)
　　增量与改正量 increment and correction
　　瞬时梯度 instantaneous gradient
　　无限冲击响应 infinite-duration impulse response
　　有限冲击响应 finite duration impulse response
　　内插 interpolating
　　晶格结构 lattice architecture
　　漏泄因数 leakage factor
　　最低次谐波 lowest-order harmonic
　　均方差 mean-square error
　　质量衰减系数 mass-attenuation coefficient
　　一套方法 methodology
　　模拟的 mimic
　　米塞斯圆柱面 Mises cylinder
　　模式分析 modal analysis
　　模带宽 modal bandwidth

系统辨识 system identification
传递迁移率 transfer mobility
变异 variance
角重叠模型 angular overlap modal

8.3 仿真软件

商业的 commercial
软件 software
    模块 module
    界面 interface
    图形用户界面 graphical user interface(GUI)
仿真 simulate
计算 calculate
3D 仿真 3D simulation
范例 paradigm
建模 model
画(布) canvas
光标 cursor
批处理 batch processing
基准 benchmark
闭合式 closed form
闭合面积 closure area
组合框 combo box
命令行界面 command line interface (CLI)
开始 commence
注解,注释 comments
超链接 hyperlink

8.3.1 图形特征

默认图表类型 default chart
青色 cyan
淡黄色 faint yellow
品红 magenta
黄褐色 tan
短划线 dashed line
点虚线 dotted
(使)染色 stain
单峰的 unimodal

8.3.2 建模

薄片理论 strip theory
连续统一体 continuum
轮廓 contour
横断面 cross-section
曲率 curvature
下摆圆角的 cutaway
迭代 iteration
膜元 membrane element
薄板,薄片 sheet
薄壁的 thin-walled
壳 shell
扁平的 tabular
桁架杆元 truss element
数据库规范化 normalised data
面向对象编程 object-oriented programming
被调用过程,被调用程序 invoked procedure
插件程序 plug-in
子程序 subroutine
端口设定 ports-settings
轮廓 outline
填补 padded
调色板 palette
工具包 toolkit
面,板 panel
下拉选项屏(单) pull-down menu
脚本 script
分开,隔开 partition
截面 section
分割的 partitioned
(电脑屏幕上)从上到下移动(资料等) scroll

轮廓 silhouette
草图 sketch
基底,基片 substrate
扫掠网格 swept meshing
视口 viewport
均衡的 well-balanced
适定的问题 well-posed problem
定义明确的 well-defined
发育良好的 well-developed

9 系统搭建

9.1 机械加工

方案 scenario
计划,方案 scheme
纲要的,图表的 schematic
后加工 postprocessing
预定尺寸 redetermined size
加工,处理 processing
副产品 by-product
轮廓 profile
计划 programme
中心的 centric
可量测性,可伸缩性 scalability
缩放比例 scaling
容许公差 allowable tolerance
余隙 clearance
空隙 interstice
以手动方式 manually
制造业 manufacturing
加工品 artifact
微细加工 microfabrication
工具 tools
 夹具 chucking
 夹子 clip
 钻头 drill
 吊架 hoist
 架空吊运车 skyhook
 管,筒,线轴 spool
 搭扣 buckle
 管夹 pipe clamp
 滚压机 roller
方法 method

透视 perspective
用针和酸类在金属板上蚀刻(图画等) etch
焊接,熔接 weld
冲压 stamping
扭 twist
电镀 electroplating
工件 workpiece
 轴衬,套管 bushing
 滚珠丝杠 ball screw shaft
 滚针轴承 needle bearing
 螺栓 bolt,threaded bolt
 长螺栓 tie-rod bolt
 波纹管 bellows
 加衬管 lined duct
 孔 orifice
 垫 pad
 钉扎点 pinning point
 销连接 pinning
 标准漏孔 reference leak
Y字形的东西 wishbone
有两边的 two-sided
∧形或∨形标志 chevron
楔形物 wedge
开槽的 grooved
旁通 bypass
端盖 end cap
边缘 fringe
平截头体 frustum
歪 skew
厚板 slab
螺线管 solenoid
尖状物 spike
螺旋形的 spiral
回弹角 springback angle
正齿轮 spur gear
酯 ester
水平的 horizontal

垂直的 perpendicular
纵向地 lengthwise
布满 line with
侧面的 lateral
金属的 metallic
中跨 midspan
周围 perimeter
外围 periphery
（合成）橡胶 rubber
橡胶硬度测试器 rubber hardness tester
斜板 skew plate
方格的 squared
负遮盖阀 underlapped valve
重量损失 weight penalty
与宽同方向地,横向地 widthwise

9.2 装 配

对齐 align
分配 allocation
轴向 axial
轴对称的 axisymmetric
方位角 azimuth
倒置 bottom-up
装配 assembly
辅助的 auxiliary
带有障板的 baffled
环氧的 epoxy
环境 environment
环境噪声 ambient noise
相等地 equally
等距的 equidistant
平衡 equilibrium
等价的 equivalent
压挤 extruding
失调 misadjustment
未对准 misalignment
台,平台 platform

树脂,松香 resin
涂抹 smear
平动,平移 translational

9.3 仪器设备

9.3.1 常规仪器

游码标尺 rider bar
千分尺 micrometer
使标准化 standardize
单元 module
调准,校准 calibration
电缆 cable
铠装线 wire harness
接线 wiring
光纤线路 fibre circuit
硬件 hardware

9.3.2 信号测量与采集

试验台仪表 test rig instrument
整形滤波器 shaping filter
滤波器 filter
抗混叠滤波器 anti-aliasing filter
假频滤波器 aliasing filter
功率放大器 power amplifier
带通滤波器 band pass filter (BPF)
低通滤波器 low pass filter
高通滤波器 high pass filter
带阻滤波器 band stop filter
宽带放大器 wide-band amplifier
宽频带滤波器 wide-band filter
检测装置,扫描设备,扫描器 scanner
转速计 tacho
滤除 filter out
线性调频 chirp

检波 demodulation

同步传输模式 synchronous transfer mode

异步传输模式 asynchronous transfer mode

同步光纤网 synchronous optical network

(使)同步,(使)同速进行 synchronize

奇偶校验 even-odd check

以太网 Ethernet

编解码器 codec

重合寄存器 coincidence register

效力 efficacy

效率 efficiency

主机 host

沙漏 hourglass

激振机 shaker

专用计算机 special-purpose computer

单机 stand-alone

10 应用领域

10.1 通　用

机械布置 machinery arrangement
机械系统 mechanical system
机械装置 mechanism
离心的 centrifugal
电动机 motor
　　电枢 armature
　　转子 rotor
润滑,加油 lubrication
发动机 engine
　　活塞 piston
　　外壳 housing
　　发动机架 engine cradle
　　发动机的排气管道 engine's exhaust pipe
　　平均自由程 mean free path
齿轮啮合 gear mesh
齿轮齿 gear teeth
齿轮系 gear train
齿廓,齿形 tooth profile
变速箱,变速器 gearbox
左舷 port
右舷 starboard
吸声衬垫 acoustic lining
剪力接合器 shear connector
腔 cavity
电磁阀 electromagnetically operated valve
柱形平衡阀 cylindrically balanced valve
张力调整 tensioning
货物(量) cargo
有效载荷 payload
隔间 compartment
协调传动 co-ordinated drive
冷却液 coolant
连接器 coupling
曲轴皮带盘 crankshaft pulley
巡航 cruise
退耦 decoupling
传动轴 drive shaft
排水干 drain
导管 duct
(机械等的)凸缘 flange
侧向传声 flanking transmission of sound
高压 high pressure
内部装饰 interior trim
层流翼 laminar flow wing
视向速度 line-of-sight velocity
液面警报器 liquid-level alarm
损耗角 loss angle
功耗因素 loss factor
装载 mount
多种的 multifold
导航 navigate
喷嘴挡板机构 nozzle-flapper mechanism
桨叶基频 fundamental blade passage frequency
在船(火车、飞机、汽车)上 on board
军械 ordnance
弹道 trajectory
方向 orientation
离地面的 overhead
操纵阀 pilot valve
管线 pipes
俯仰 pitching
装在枢轴上的 pivoted
传动系 power train
桨扇发动机 prop-fan-engines

推进 propulsion
原型的 prototypical
原型机制造 prototyping
行李架 rack
乘坐品质控制 ride quality control
机器人技术 robotics
舵 rudder
配套齐全的 self-contained
剪力接合器 shear connector
剪力自锁 shear locking
短程的 short-haul
时空的 spatio-temporal
烟囱管 stove-pipe
水陆运输 surface transport
触觉的 tactile
运动目标移动痕迹 tails of moving targets
紊流边界层 turbulent boundary layer (TBL)
紊流 turbulent fluid
控制油、气流的阀门 throttle
肘节角 toggle angle
涡轮轴发动机 turbo shaft engine
涡轮增压柴油机 turbocharged diesel engine
啸声 whistling
低音用扩音器 woofer
调频质量阻尼器 tuned mass damper
音圈型调节器 voice-coil type actuator
阻尼器 dashpot
缓冲器 buffer
检测质量 proof mass
反共振,防共振,电流谐振 anti-resonance

10.2 航空航天

航空航天工业 aerospace
航空电子学的 avionic
天文学 astronomy
制导 guidance
陀螺仪,回转仪 gyroscope
飞机引擎 aeroengine
叶片流道 blade passage
(船、飞机、导弹等的)后体,尾部 afterbody
机载的 airborne
航空电子学的 avionic
螺旋桨 propeller
直升机 helicopter
旋翼飞机 rotorcraft
固定翼 fixed wing
(飞机的)机身 fuselage
舷门 gangway
侧翼 flank
客舱 passenger cabin
巡航 cabin
乘坐品质控制 ride quality control
噪声消除 noise cancellation
重量损失 weight penalty
安全分离装置 breakaway device
(使)升空 levitation

10.3 车辆

车辆 vehicle
车的 vehicular
出租车 cab
长途客运汽车 coach
槽形轨,电车轨 girder rail
货车 truck
轨道 track

公路 highway
高速公路 motorway
驾驶舱 cabin
车轴 axle
断开器,分离器 decoupler
毂 hub
悬架刚度 suspension stiffness
转向构架 bogie frame
变速箱,变速器 gearbox
(齿)轮系,传动机构,齿轮传动链 gear train
转向架中心销 bogie pivot
安全分离装置,安全脱钩装置,保险装置 breakaway device
自动防故障装置 fail-safe
把(铁路货车等)调到另一轨道上 shunt
制动力 braking force
底盘 chassis
离合器 clutch
轨枕间距 sleeper pitch

10.4 舰　　船

船 vessel
偏航 yaw
运河 canal
甲板纵桁 deck girder
船桅 mast
声呐装置,声呐系统 sonar
磁发电机 magneto
陀螺仪,回转仪 gyroscope
排水管 drain
输送管,导管 duct
船头装货起重机 bow charging crane
感音水雷 acoustic mine
船体 hull
潜艇 submarine
隔离壁 bulkhead
驼背 humpback
(船的)龙骨 keel
节,海里/小时 knot
海洋的 marine
海军的 naval
近海的 offshore
海上航道 seaway
声呐装置 sonar
秘密行动 stealth
艉轴管轴封泵 stern tube sealing oil pump
在水中的 submerged
超空蚀水流 supercavitating flow
舵手室 wheelhouse

10.5 建　筑　物

地震的,由地震引起的 seismic
钢筋混凝土 reinforced concrete
预应力的(混凝土) prestressed
房间 chamber
盖瓦 tiling
变压器 transformer

10.6 家用电器、消费类电子

(尤指颅骨)基部的,底部的 basilar
螺旋形的 spiral
顶盖的,覆膜的 tectorial
耳蜗的 cochlear
耳蜗 cochlea
外耳 concha
耳罩 ear muff
头挂听筒 head phone
靠枕 headrest
耳机 headset

耳听力发射处理器 otoacoustic emission processor

听筒 receiver

个人数码娱乐产品 personal digital entertainment products

球棒 bat

附录 A 英汉对照

(sigma) summation of 总和
3D simulation 3D 仿真
A/D converter A/D 转换器
abiding 永久的
absolute convergence 绝对收敛
absolute error 绝对误差
absolute extreme values 绝对极值
absolute maximum and minimum 绝对极大与极小
absolute stability 绝对稳定
absolute temperature 绝对温度
absolute value function 绝对值函数
absolute value 绝对值
absorption coefficient 吸收系数
AC power supply 交流电源
acceleration resonance frequency 加速度共振频率
acceleration sensors 加速度传感器
acceleration 加速度
acceptance region 接受区域
accuracy 准确度
acoustic impedance 声阻抗
acoustic lining 吸声衬垫
acoustic mine 感音水雷
acoustic mode 声模态
acoustic potential energy 声势能
acoustic resonance 声共振
acoustic transmission coefficient 声透射系数,传声系数
acoustic 声学
active power 有功功率
active sensor 主动传感器
active 主动
actuation 作动
actuator materials 作动器材料
actuator 作动器
adaptation 自适应
adaptive control 自适应控制
addition theorem 加法定理
addition 加
adiabatic compression 绝热压缩
adiabatic 绝热的
admittance 导纳
aeroacoustics 空气声学
aerodynamic 空气动力学的
aeroelastic 空气弹性变形的
aeroengine 飞机引擎
aerospace 航空航天工业
afterbody(船、飞机、导弹等的)后体,尾部
airborne 机载的
algebraical 代数的
aliasing filter 混叠滤波器
align 对齐
allocation 分配
allowable tolerance 容许公差
alloy families that exhibit shape memory effect 显示形状记忆效应的合金系列
alternating current(AC)交流电
ambient noise 环境噪声
American Standard Code for Information Interchange ASCII 代码
Amperes's Law 安倍定律
ampersand & 符号
amplification circuit 放大电路
amplifier 放大器
amplify component 放大元件
amplitude-frequency characteristics 幅频特性
amplitude 振幅
analog integrated circuit 模拟集成电路

analogue interpolation 类比插值法
analog 模拟
anechoic room 消声室
anechoic 无回声的
angle 角
angular acceleration 角加速度
angular frequency 角频率
angular overlap modal 角重叠模型
anisotropic 各向异性的
annihilating polynomial 化零多项式
ant colony algorithm 蚁群算法
anti-aliasing filter 抗混叠滤波器
antiderivative 反导数
antinode 波腹
antiphase 反相
antiresonance 反共振
anti-resonance 反共振,防共振,电流谐振
anti-symmetric matrix/skew-symmetric matrix 反对称矩阵
aperiodic 非周期
apostrophe 撇号
apparent power 视在功率
approximate integration 近似积分
arbitrary polynomial 任意多项式
area 面积
arithmetic 四则运算
armature 电枢
arrow 箭头
artifact 加工品
as(proportion) 成比例
assembly 装配
asterisk 星号
astronomy 天文学
asymptote 渐近线
asymptotic line 渐近线
asymptotically efficient estimate 渐近有效估计
asymptotically stable 渐近稳定
asynchronous transfer mode 异步传输模式
attenuation vibration 衰减振动
audiology 听力学
austenite 奥氏体
auto-correlation 自相关
automatic control system 自动控制系统
automatic control 自动控制
autonomous sensing 自感知
autonomous system 自治系统
autoregressive 自回归
auxiliary 辅助的
averaging power 平均功率
avionic 航空电子学的
A-weighted sound pressure level A[计权]声[压]级权
axes 坐标轴
axial 轴向
axiom 公理
axisymmetric 轴对称的
axle 车轴
azimuth 方位角
backlash 间隙
backslash 反斜线转义符,有时表示转义符或续行符
baffled 带有障板的
ball screw 滚珠丝杠
band-stop filter 带阻滤波器
band pass filter(BPF) 带通滤波器
bandwidth 带宽
barotropic 正压
base units 基本单位
base 基
basilar(尤指颅骨)基部的,底部的
batch processing 批处理
bat 球棒
bayes formula 贝叶斯公式,逆概率

公式
 beats 节拍
 because 因为
 bellows 波纹管
 benchmark 基准
 bending deflection 弯曲变形
 bending elasticity 弯曲弹性
 bending modulus 弯曲模量
 bending moment 弯矩
 bending stiffness 弯曲刚度
 bending strain 弯曲应变
 Bernoulli constant 伯努利常数
 Bernoulli distribution 伯努利分布
 Bernoulli law of large numbers 伯努利大数定律
 best estimate 最佳估计
 bias circuit 偏置电路
 bimetallic strips 双金属条
 bimorph 双压电晶片元件
 binary 二进制的
 Bingham body 宾汉塑性体
 Bingham model 宾汉模型
 Bingham plastic fluid 宾汉塑性流体
 binomial distribution 二项分布/伯努利分布
 binomial series 二项级数
 bioacoustics 生物声学
 biological 生物
 bipolar integrated circuit 双极型集成电路
 black box model 黑箱模型
 blade passage 叶片流道
 block diagram 方块图
 block Toeplitz system 分块 Toeplitz 系统
 blocked force 阻塞力
 Bode diagram 伯德图
 body force 体力

bogie frame 转向构架
bogie pivot 转向架中心销
bolt 螺栓
bottom-up 倒置
boundary condition 边界条件
boundary stiffness matrix 边界刚度矩阵
bow charging crane 船头装货起重机
brackets 括号
braking force 制动力
branch current method 支路电流法
branch 支路
breakaway device 安全分离装置,安全脱钩装置,保险装置
bridge rectifier 桥式整流器
broad band 宽带
buckle 搭扣
buffer 缓冲器
bulk(巨大)物体
bulkhead 隔离壁
buoyancy 浮力
bushing 轴衬,套管
Butterworth filter 巴特沃斯滤波器
bypass 旁通
byte 字节
cabin 驾驶舱
cable 电缆
cab 出租车
calculate 计算
calculus 微积分
calibrate 使标准化
calibration 调准,校准
canal 运河
cancel out 抵消
cancellation point 对消点
canonical 标准的
cantilever 悬臂
canvas 画(布)

— 47 —

capacitance change 电容变化
capacitance 电容
capacitive sensors 电容传感器
capacitive 容性
capacity 容量
cargo 货物(量)
Cartesian coordinate system 笛卡儿坐标系
Cartesian coordinates 笛卡儿坐标
cascaded 级联的
Cauch's mean value theorem 柯西均值定理
Cauchy distribution 柯西分布
causal system 因果系统
causal 因果关系
cavity 腔
Cayley-Hamilton theorem 凯莱－哈密顿定理
Celsius system 摄氏度
centric 中心的
centrifugal 离心的
centripetal force 离心力
cepstral analysis 倒谱分析
cepstral distance 倒谱距离
ceramic-glass 陶瓷玻璃
ceramic-polymer 陶瓷复合材料
ceramics 陶瓷
certain event 必然事件
certain phenomenon 必然现象
chamber 房间
change-of-state temperature measurement 状态变化温度测量
chaotic 混沌的
characteristic curve 特性曲线
characteristic equation 特征方程
characteristic length 特征长度
characteristic polynomial 特征多项式
characterizing definition 特征定义

charge 电荷
chassis 底盘
Chebyshev inequality 切比雪夫不等式
chemical 化学
chevron ∧形或∨形标志
chirp 线性调频
chucking 夹具
circle 圆
circuit 电路
circular 环形的
circumference 圆周
circumferential 圆周的
civilian 平民
clamped 固支
classic control theory 经典控制理论
classic input signal 典型输入信号
classical probabilistic model 古典概率模型
classical probability 古典概率
clearance 余隙
clip 夹子
close brace,close curly 右花括号
close bracket 右方括号
close parenthesis 右圆括号
closed form 闭合式
closed interval 封闭区间
closed-loop control 闭环控制
closed-loop transfer function 闭环传递函数
close-loop control 闭环控制
closure area 闭合面积
clustering 聚类
clutch 离合器
coach 长途客运汽车
cochlear 耳蜗的
cochlea 耳蜗
codec 编解码器
coder,encoder 编码器

coding 编码
coefficient 系数
coercive force 抗磁力
coherence 一致
coherent wave 相干波
coil resistance 线圈阻尼
coincidence frequency 相干频率
coincidence register 重合寄存器
collocated 同位置的
collocation method 同位置配置法
colon 冒号
combination circuit 组合电路
combo box 组合框
command line interface(CLI) 命令行界面
comma 逗号
commence 开始
comments 注解，注释
comment 注释符
commercial 商业的
common-mode rejection ratio(CMRR) 共模抑制比
common-mode signals 共模信号
communications 通信
compact 紧凑的
compare component 比较元件
compartment 隔间
compatible equations 相容方程组
compatible norm 相容范数
compensation 补偿
complementary 互补的
complex admittor 复导纳
complex domain method 复域法
complex impedance 复阻抗
complex power 复功率
compliance 顺应性
compressible fluids 可压缩流体
compression 压缩

computational fluid dynamic(CFD) 计算流体力学 CFD
computationally infeasible 计算上不可能的
computation 计算
concave function 凹函数
concave 凹的
concentrated force 集中力
concha 外耳
condenser microphones 电容式麦克风
condition number 条件数
conditional distribution 条件分布
conditional probability distribution 条件概率分布
conditional probability density 条件概率密度
conditional probability 条件概率
conductor 导体
cone 圆锥
confidence coefficient 置信系数
confidence interval 置信区间
confidence level 置信水平
confidence lower limit 置信下限
confidence upper limit 置信上限
conformal array 共形阵
conical 圆锥形的
conjugate 共轭的
consistent norm 相容范数
consistent stability 一致稳定
consistent units 一致性量纲
constant-value system 恒值系统
constant function 常数函数
constant 常量
constituent 成分
constrain 约束
constructive interference 建设性干涉
contemplated 预期的
continuity at a point 在一点处之连

续性
    continuity from the left 左连续
    continuity from the right 右连续
    continuity of a function 函数之连续性
    continuity on an interval 在区间之连续性
    continuous（time）signal 连续（时间）信号
    continuous function 连续函数
    continuous random variable 连续随机变量
    continuous signal 连续信号
    continuous system 连续系统
    continuum 连续统一体
    contour integral 围道积分
    contour 轮廓
    control moment 控制力矩
    control object 控制目标
    control signal 控制信号
    control system 控制系统
    controllability decomposition 能控性分解
    controllability 能控性
    controllable canonical form 可控标准型
    controlled source 受控源
    convective 对流的
    converged 收敛的
    convergence in probability 依概率收敛
    convergence 收敛
    convergence 收敛性
    convergent sequence 收敛数列
    convergent series 收敛级数
    conversation of momentum 动量守恒
    converse piezoelectric effect 逆压电效应
    convex function 凸函数
    convex 凸面的
    convolution 卷积
    coolant 冷却液
    coordinate axes 坐标轴

coordinate planes 坐标平面
co-ordinated drive 协调传动
coordinate 坐标
copper-aluminum-nickel 铜－铝－镍
copper-zinc-aluminum 铜－锌－铝
corner frequency 转角频率
correction point 校正元件
correlation coefficient 相关系数
corrosive 腐蚀
cosine function 余弦函数
cost function 性能函数
cost-effective 有成本效益的
coulomb friction 干摩擦
coulomb 电量
counteract 对抗；抵消
counter 计数器
coupled pole 复合杆
coupled system 耦合系统
coupling 连接器
covariance matrix 协方差矩阵
covariance 协方差
crankshaft pulley 曲轴皮带盘
crest 顶
critical damp system 临界阻尼系统
critical point 临界点
critical stable 临界稳定
cross spectral density 互谱密度
cross-correlation 互相关
cross-section 横断面
passenger cabin cruise 客舱
cruise 巡航
crushable 可压碎的
crystallographic 晶体的
cubic function 三次函数
cumbersome 沉重的
Curie temperature 居里温度
Curie-Weiss law 居里·韦斯定律
current change 电流变化

current source 电流源
current 电
current 电流
cursor 光标
curvature 曲率
curve 曲线
cut off distortion 截止失真
cutaway 下摆圆角的
cut-off angular frequency 截止角频率
cutoff frequency 截止频率
cut-on frequency 截止频率
cyan 青色
cyclic 周期的
cylinder 圆柱
cylindrical coordinate 柱面坐标
cylindrically balanced valve 柱形平衡阀
D/A converter D/A 转换器
dashed line 短划线
dashpot 阻尼器
dash 破折号
data reduction 数据缩减
DC power supply 直流电源
De Moivre-Laplace theorem 棣莫弗－拉普拉斯定理
dead zone characteristics 死区特性
dead zone 死区
decibel 分贝
decimals 小数
decimal 十进制的
deck girder 甲板纵桁
decoder 译码器
decoupler 断开器，分离器
decoupling 退耦
decreasing function 递减函数
decreasing sequence 递减数列
default chart 默认图表类型
definite integral 定积分
degree of a polynomial 多项式之次数

degree of confidence 置信度
degree of freedom 自由度
degree 度
delay time 延时
deleterious 有害的
demodulation 检波
denominator 分母
density of distribution 分布密度
density 密度
dependence relation 相关关系
dependent variable 因变量
depth negative feedback 深度负反馈
derivative 导数
derived units 导出单位
description function method 描述函数法
designate component 给定元件
de-stabilize 降低稳定性
destructive interference 破坏性干涉
detection of trigger events 触发事件的检测
deteriorate 恶化，变坏
determinant factor 行列式因子
determinant 行列式
deterministic signal 确定性信号
detriment 损害，伤害
diagonal matrix 对角矩阵
diagonalization 对角化
diagonalizable 可对角化
diamond 菱形
diaphragm 隔膜
dielectric 电介质
difference equation 差分方程
difference 差
differentiable function 可导函数
differential equation 微分方程
differential mode signal 差模信号
diffraction 衍射
diffuse acoustic filed 扩散声场

digital circuits 数字电路
digital control system 数字控制系统
digital integrated circuit 数字集成电路
digital signal 数字信号
digital 数字
dilatant substances 缓释物质
dilatation rate 扩张率
dimensionless 无量纲的
dimension 维度
diode 二极管
dipoles 偶极子
direct current (DC) 直流电
direct piezoelectric effect 直接压电效应
direct sound field 直达声场
directional characteristics of radiation 辐射的指向特性
directional derivatives 方向导数
directivity factor 方向性因子
directivity 方向性
discipline 学科
discontinuity 不连续性
discrepancy 差异
discrete random variables 离散随机变量
discrete signal 离散信号
discrete system 离散系统
discrete-time system 离散时间系统
displacement resonance frequency 位移共振频率
displacement sensor 位移传感器
displacement 位移
distortion 失真
distributed constant (parameter) circuit 分布参数电路
distributed control 分布式控制
distribution function 分布函数
disturbance 干扰
divide 除

divide 除号
dominant closed-loop poles 闭环系统的主导极点
Doppler effect 多普勒效应
dots 省略号
dotted 点虚线
dot 点
double integral 二重积分
double quotation marks 双引号
doubling rectifier 倍压整流
downstream 下游
drain 排水管
drill 钻头
drive shaft 传动轴
duality 二元性
dual 二重的
duct 输送管,导管
duct 导管
dynamic characteristics 动态特性
dynamic model 动态模型
dynamic state 动态
dynamic system 动态系统
dynamometer 测功计
ear hearing transmitter 耳听力发射处理器
ear muff 耳机
eardrum 鼓膜
effective sound pressure 有效声压
efficacy 效力
efficiency of estimator 估计量的有效性
efficiency 效率
eigenvalue 特征值
eigenvector 特征向量
elastic control area 弹性控制区
elastic modulus 弹性模量
elastic resistance 弹性抗
elasticity 弹力
elastodynamics 弹性动力学

elastomeric 人造橡胶的
elastomer 人造橡胶
elasto 弹塑性
electric actuator 电动作动器
electric potential 电位
electric power 电功率
electrical field 电场
electrical 电
electrode 电极
electro-dynamometer 电功率计
electromagnetic actuator 电磁作动器
electromagnetically operated valve 电磁阀
electromagnetic 电磁
electromechanical 机电的
electron hole pair 电子–空穴对
electroplating 电镀
electrorheological & magnetorheological fluids 电流变液和磁流变液
electrostatic 静电的
elementary divisor 初等因子
elementary event 基本事件
eletrocnic tube 电子管
ellipse 椭圆
ellipsis 省略号
ellipsoid 椭圆体
empirical distribution function 经验分布函数
enclosure space 封闭空间
enclosure 围场
encoders 编码器
end cap 端盖
endogenous 内长的
energy conservation 能量守恒
energy harvest 能量收集
energy storage components 储能元件
energy 能量
engine cradle 发动机架

engine's exhaust pipe 发动机的排气管道
engine 发动机
ensor characteristics 传感器特性
enthalpy 焓
entropy 熵
environment 环境
epicycloid 外摆线
epoxy 环氧的
equally likely event 等可能事件
equally 相等地
equals 等于
equation 方程式
equidistant 等距的
equilateral 等面的
equilibrium position 平衡位置
equilibrium state 平衡状态
equilibrium 平衡
equivalent 等价的
ergodicity 各态历经
ergodic 遍历性的
error transfer function 误差传递函数
error 误差
ester 酯
estimation of population variance 总体方差的估计
estimator 估计量
etch 用针和酸类在金属板上蚀刻（图画等）
Ethernet 以太网
Euler's equation of motion 欧拉运动方程
evanescent 迅速消失遗忘的
even function 偶函数
even-odd check 奇偶校验
evolve 发展
excitation frequency 激励频率
excitation 激励

exclamation mark 感叹号(英式英语)
exclamation point 感叹号(美式英语)
execute component 执行元件
exhaustive search 穷举搜索
exogenous 外生的
expectation value 期望值
expected valued 期望值
experiment 试验
expert control 专家控制
explosive 爆炸的
exponential distribution density function 指数分布密度函数
exponential distribution 指数分布
exponential function 指数函数
exponentially stable 指数式稳定
extension 延伸
extraneous 外部的
extrapolate 推算
extrema 极值(extremum 的复数)
extreme value theorem 极值定理
extreme value 极值
extrinsic 外在的
extruding 压挤
extrusion 挤压
factorial 阶乘
fail-safe 自动防故障装置
faint yellow 淡黄色
far field 远场
Faraday's Law of Induction 法拉第诱导定律
far-field approximation 远场近似
fault tolerant control 容错控制
feedback amplifier 反馈放大器
feedback channel 反馈通道
feedback compensation 反馈补偿
feedback control 反馈控制
feedback gain matrix 反馈增益矩阵
feedforward control 前馈控制

ferrite 铁酸盐
ferromagnetic 铁磁的
fiberglass 玻璃纤维
fibre circuit 光纤线路
fibre-optic strain gauges 光纤应变计
field effect transistor 场效应管
filament 细丝
film 薄膜
filter out 滤除
filter 滤波器
finite duration impulse response 有限冲击响应
first derivative test 一阶导数试验法
first octant 第一卦限
fixed wing 固定翼
flammable 易燃的
flange (机械等的)凸缘
flanking transmission of sound 侧向传声
flank 侧翼
flexural rigidity 抗挠刚度
flip-flop 触发器
flowchart 流程图
fluid dynamics 流体动力学
fluid expansion devices 流体膨胀装置
fluidic 流体的
fluid 流体
fluoride 氟化物
flux 连续的改变
foam 泡沫
force control area 力阻控制区
force gauge 测力计
force generation 力生成
force impedance 力阻抗
force resistance 力阻
force sensor 力传感器
force 力
forcing frequency 扰动频率
formula of total probability 全概率公式、

全概率定理
    forward channel 正向通道
    Fourier transform 傅里叶变换
    fraction 分式
    framework 框架
    free body 自由体
    free field correction 自由场校正
    free field 自由场
    free vibration 自由振动
    free-body diagram 受力图
    free-field method 自由场法
    frequency bin 频率窗口
    frequency domain method 频域法
    frequency number 频率数
    frequency of occurrence 出现次数
    frequency response function 频率响应函数
    frequency-response method 频率响应方法
    frequency 频率
    friction characteristics 摩擦特性
    friction force 摩擦力
    fringe 边缘
    frustum 平截头体
    F-test F 检验
    full stop 句号
    full transmission 全透射
    full-adder 全加器
    full-rank decomposition 满秩分解
    full-wave rectification 全波整流
    fully-compensated operational amplifier 全补偿运算放大器
    functional materials 功能材料
    fundamental blade passage frequency 桨叶基频
    fundamental frequency 基频
    fused 装有保险丝的
    fuselage (飞机的)机身
    fuzzy control 模糊控制
    gain margin 幅值裕度
    gangway 舷门
    gap characteristics 间隙特性
    gas molecules 气体分子
    gauge length 标距长度
    gear mesh 齿轮啮合
    gear train (齿)轮系,传动机构,齿轮传动链
    gearbox 变速箱,变速器
    General Conference of Weights & Measures (CGPM) 国际计量大会(CGPM)
    general solution 通解
    generalized root locus 广义根轨迹
    genetic algorithm 遗传算法
    geometric probability 几何概率
    geometric series 几何级数
    geometry 几何
    geotechnical engineering 岩土工程
    global 全局的
    gradient 梯度
    graphical user interface (GUI) 图形用户界面
    graphite 石墨的
    gravity 重力
    gray box model 灰箱模型
    grazing incidence 掠入射
    grider rail 槽形轨,电车轨
    grooved 开槽的
    guidance 制导
    gyroscope 陀螺仪,回转仪
    half adder 半加器
    half wave rectification 半波整流
    Hankel function 汉克尔函数
    hard boundary 硬边界
    hardware 硬件
    harmonic frequency generation 谐频发生
    harmonic frequency 谐频

harmonic series 调和级数
harmonic wave 简谐波
head phone 头挂听筒
headrest 靠枕
headset 耳机
hearing loss 听力损害
heat diffusivity 热扩散率
heat flux density 热通量密度
helicopter 直升机
helix 螺旋线
helmholtz resonator 亥姆霍兹共振器
hence 所以
Hermitian quadratic form 厄米二次型
hermitian 厄密共轭
Hessian matrix 海赛矩阵
hexadecimal 十六进制的
hexahedral 六面体的
high pass filter 高通滤波器
high pressure 高压
higher derivative 高阶导数
higher-order 高阶
highway 公路
hinged 铰支
hoist 吊架
homogeneity test for variances 方差齐性的检验
homogeneous principle 齐次原理
homogeneous 同性质的
homogeneous 同质
honeycomb 蜂窝
Hooke's Law 胡克定律
horizontal asymptote 水平渐近线
horizontal line 水平线
horizontal 水平的
host 主机
hourglass 沙漏
housing 外壳
hub 毂

hull 船体
humpback 驼背
Hurwitz's stability criterion 赫尔维茨稳定判据
hybrid integrated circuit 混合集成电路
hybrid 杂交生成的生物体
hydraulic actuator 液压作动器
hydraulic 液压的
hydromodulus 流量模数
hyper boloid 双曲面
hyperbola 双曲线
hyperbolic 双曲线的
hypergeometric distribution 超几何分布
hyperlink 超链接
hypersonic 远超过音速的
hyphen 连字号
hypothesis testing 假设检验
hypothesis 假设
hypothetical 假设的
hysteresis 磁滞
ideal fluid 理想流体
identity matrix 单位阵
identity 同一性
ill-conditioned 病态
ill-posed problem 不适定问题
image number 虚数
impedance 阻抗
impetus 推动
implication of events 事件的包含
implicit function 隐函数
impose 施加
impossible event 不可能事件
improper integral 瑕积分
impulse function 脉冲函数
impulse response function 脉冲响应函数
impurity semiconductors 杂质半导体
incident normal 入射法线
incident wave 入射波

incompatible events 互不相容事件，互斥事件
incompressible fluid 不可压缩流体
increment and correction 增量与改正量
indefinite integral 不定积分
independent event 独立事件
independent experiment 独立实验
independent Levy-Lindberg theorem 列维－林德伯格定理、独立同分布中心极限定理
independent random event 独立随机事件
independent random variables 独立随机变量
independent variable 自变量
indeterminacy 不确定性
individual 个体
indoor reverberation 室内混响
induced 感应的
inductance 电感
inductive 感性
inelastic buckling 非弹性屈曲
inertial force 惯性力
inertial frame 惯性系统
inertial sensor 惯性传感器
infinite duration impulse response 无限冲击响应
infinite series 无穷级数
infinitesimal 极微小的
infinity 无限大
inflation 膨胀
inflection point 拐点
information 信息
infrared devices 红外设备
infrasonic wave 次声波
inhomogeneity 不均匀性
initial value theorem 初值定理
inner product spaces 内积空间
inner product 内积

in-phase 同步
input impedance 输入阻抗
input resistance 输入电阻
instantaneous gradient 瞬时梯度
instantaneous power 瞬时功率
instantaneous sound pressure 瞬时声压
insulation 隔离
insulator 绝缘体
integer multiple 整数倍数
integer 整数
integral enclosure 整体密闭罩
integral transform 积分变换
integrated circuit 集成电路
integrated operational amplifier 集成运算放大器
integration by part 分部积分法
integration coupling of sensing and computation at the chip level 传感和计算在芯片级上耦合
integration sequence 整合序列
integrator 积分器
intelligent control theory 智能控制理论
intensity amplification 强度放大
interaction 相互作用
intercepts 截距
interchangeably 可交换地
interconnect 互相连接
interdisciplinary 各学科间的
interface 界面
interfacing 界面
interference 干涉
interim 暂时的
interior trim 内部装饰
interleave 交织
intermediate value of theorem 中值定理
intermediate 中间的
intermittent noise 间歇性噪声
intermittent 间歇的

internal friction 内摩擦
internal modal control(IMC) 内模控制
interpolating 内插
intersection events 交事件
intersection of 交,通集
interstice 空隙
interval estimation 区间估计
interval of convergence 收敛区间
interval 区间
intrinsic nonlinearity 本质非线性
intrinsic semiconductor 本征半导体
intrinsic 固有的
invariant factor 不变因子
invasive 有的扩散危害
inverse function 反函数
inverse matrix 逆矩阵
inverse trigonometric function 反三角函数
invertible 可逆
invoked procedure 被调用过程,被调用程序
iron-manganese-silicon 铁-锰-硅
irrational number 无理数
irregularities 不齐
irregular 不对称的
irrotational flow 无旋流
is approximately equal to 约等于
is divided by 除
is equal to or approximately equal to 等于或约等于
is equivalent to 全等于
is less than or equal to 小于或等于
is less than 小于
is more than or equal to 大于或等于
is more than 大于
is multiplied by 乘
is not equal to 不等于
is not less than 不小于
is not more than 不大于
isentropic 等熵
isolation amplifier 隔离放大器
isometric 等大的
isosceles 等腰的
isothermal 等温
isotropic 各向同性
iteration 迭代
I-beam I型梁
joint distribution function 联合分布函数
joint probability variable 联合概率密度
Jordan canonical form 约当标准型
Julie criterion 朱利判据
Karnaugh map 卡诺图
keel (船的)龙骨
Khinchine law of large numbers 辛钦大数定律
kinematic viscosity 动黏滞率,动黏度
kinematically 运动学上地
Kirchhoff's law 基尔霍夫定律
knot 节,海里/小时
Kronecker symbol 克罗内克符号
lag compensator 滞后补偿
lag-lead compensation 滞后-超前补偿
laminae 薄片
laminar flow wing 层流翼
laminar 由薄片或层状体组成的
laminated (将薄片砌合在一起)制成(材料)
Laplace inverse transform 拉普拉斯反变换
Laplace transform 拉普拉斯变换
large-scale asymptotic stability 大范围渐近稳定
large-scale complex system control 大系统复杂系统控制
lateral 侧面的
lattice architecture 晶格结构

law of conditional distribution 条件分布律

law of large numbers 大数定律

law of large numbers and central limit Theorem 大数定律及中心极限定理

law of marginal distribution 边缘分布律

law of probability distribution 概率分布律

lead compensator 超前补偿

lead magnesium niobate (PMN) 铅铌酸盐

lead meta niobate (LMN) 铌酸铅

lead titanate (LT) 钛酸铅

lead zirconate titanate (PZT) 锆钛酸铅

leakage factor 漏泄因数

least significant bit (LSB) 最低有效位

least-squares solution 最小二乘解

left-hand derivative 左导数

left-hand limit 左极限

left-sided sequence 左边序列

lemma 引理

lengthwise 纵向地

length 长度

levitation （使）升空

Liapunov theorem 李亚普诺夫定理

lightweight 轻量的

light 光

likelihood equation 似然方程

likelihood function 似然函数

limit cycle 极限环

limiting circuit 限幅电路

limp 易弯的

line force 线力

line with 布满

linear algebra 线性代数

linear approximation 线性近似

linear dependence 线性相关

linear dimension 线性尺寸

linear equations 线性方程组

linear equation 线性方程式

linear function 线性函数

linear independence 线性无关

linear map 线性映射

linear regression 线性回归

linear space 线性空间

linear system 线性系统

linear time invariant system 线性时不变系统

linear time variant system 线性时变系统

linear transform 线性变换

linear variable differential transformer 线性可变差动变压器

linearised wave equation 线性波动方程

linearity 线性度

linearization of nonlinear system 非线性系统的线性化

linearization 线性化

lined duct 加衬管

line-of-sight velocity 视向速度

line-to-line voltage 线电压

line 线

liquid-level alarm 液面警报器

load 负载

local extreme values 局部极值

local maximum and minimum values 局部极大值与极小值

local 局部的

logarithmic function 对数函数

logarithmic plot 对数图

logarithm 对数

logic control 逻辑控制

logic gates 门电路

logical function 逻辑函数

logical variable 逻辑变量

longitudinal waves 纵波

loop current method 回路电流法
loop 回路
loss angle 损耗角
loss factor 功耗因素
loudness level 响度级
low pass filter 低通滤波器
lowest-order harmonic 最低次谐波
lubrication 润滑,加油
lumped-parameter circuit 集中参数电路
Lyapunov's first approach 李亚普诺夫第一方法
Lyapunov's second approach 李亚普诺夫第二方法
machinery arrangement 机械布置
macroscopically 肉眼可见的
magenta 品红
magnetic field 磁场
magnetic flux 磁通
magnetic 磁
magnetostrictive materials 磁致伸缩材料
magneto 磁发电机
magnitude 振幅
manually 以手动方式
manufacturing 制造业
mapping theorem 映射理论
marginal density 边缘密度
marginal distribution function 边缘分布函数
marginal distribution 边缘分布
marginal probability density 边缘概率密度
margin 差数
marine 海洋的
martensite 马氏体
mass conservation 质量守恒
mass control area 质量控制区

mass resistance 质量抗
mass-attenuation coefficient 质量衰减系数
massive 大而重的
mass 质量
master-slave RS flip-flop 主从 RS 触发器
mast 船桅
material density 材料密度
material properties 物料性质
material stype 材料类型
mathematical expectation 数学期望、均值
mathematical model 数学模型
matrix exponential 矩阵指数
matrix 矩阵
maturity 成熟
maximum and minimum values 极大与极小值
maximum likelihood estimation 极大似然估计
maximum likelihood estimator 极大似然估计量
maximum likelihood method 极大似然法
maximum overshoot 最大超调
mean free path 平均自由程
mean square deviation 均方差
mean value theorem 均值定理
mean–square error 均方差
measure component 测量元件
mechanica motion 机械运动
mechanical impedance 机械阻抗
mechanical properties 机械性能
mechanical system 机械系统
mechanical wave 机械波
mechanical 机械
mechanics 力学

mechanism 机械装置
meeting point 会合点
membrane element 膜元
mesh current method 网孔电流法
mesh 网孔
Meson function 梅森公式
metallic 金属的
methodology 一套方法
method 方法
micro electromechanical systems(mems) 微机电系统
microfabrication 微细加工
micrometer 千分尺
midspan 中跨
mimic 模拟的
miniature 小型的
miniaturise 小型化
miniaturization 使小型化
minimal polynomial 最小多项式
minimax 极小化极大算法
minimum least-squares solution 极小最小二乘解
minimum norm solution 极小范数解
minimum polynomial 最小多项式
minimum-phase system 最小相系统
minimum 极小值
minima(minimum 的复数)
minus 减,负
minute 分
misadjustment 失调
misalignment 未对准
Mises cylinder 米塞斯圆柱面
modal analysis 模式分析
modal bandwidth 模带宽
mode amplitude 模振幅
mode shape 振形
model 建模
modern control theory 现代控制理论
modify 修改
modularity 模块性
modulation 调制
module 模块
moduli space 模空间
modulus 系数
mold flow 模内流动性
mole 物质的量
moment estimation 矩法估计
moment of inertia 惯性矩
momentum conservation 动量守恒
momentum 动量
moment 力矩
monopole 单极子
monotonically damped stable point 单调阻尼稳定点
most significant bit(MSB) 最高有效位
motorway 高速公路
motor 电动机
mount 装载
multi channel control 多通道控制
multi-input/multi-output(MIMO) 多输入多输出
multi-dimensional 多维
multifold 多种的
multi-layer 多层
multi-level control system 多级控制系统
multi-modal 多模式
multiple integrals 多重积分
multiple roots 重根
multiplication of probabilities 概率乘法
multiplier 乘子
multiply 乘
multipoles 多极子
multiscale analysis 多尺度分析
multivariate random variables and distributions 多维随机变量及其分布

multivibrator 多谐振荡器
mutual inductance 互感
mutual radiation impedance 互辐射阻抗
mutually exclusive events 互不相容事件,互斥事件
mutually inverse 互逆的
natural exponential function 自然指数函数
natural logarithm function 自然对数函数
natural number 自然数
naval 海军的
navigate 导航
n-dimensional distribution function n 维分布函数
n-dimensional probability distribution n 维概率分布
n-dimensional random variable n 维随机变量
n-dimensional sample space n 维样本空间
near field 近场
necessary and sufficient condition 充分必要条件
needle bearing 滚针轴承
neoprene 氯丁(二烯)橡胶
network protocol standardization 网络协议标准化
neural network control 神经网络
neutralizer 中性化器
Newton's first law 牛顿第一运动定律
Newton's second law 牛顿第二运动定律
Newton's third law 牛顿第三运动定律
nickel-titanium(nitinol) 镍钛(镍钛诺)
nickel 镍
nitinol 镍钛诺(镍和钛的合金)
nodal line 节线
nodal 节的
node 节点
noise cancellation 噪声消除
nomenclature (某一学科的)术语,专门名称
non-affiliated direction 非关联方向
noncasual system 非因果系统
non-collinear vectors 非共线矢量
non-contact 非接触
non-dimensional 无量纲的
nonhomogeneous 非齐(次,性)的
nonlinear control 非线性控制
nonlinear distortion 非线性失真
nonlinear system 非线性系统
nonminimum-phase system 非最小相系统
nonparametric model 非参数模型
non-recursive 非递归的
non-singular 非奇异的
non-stationary 非平稳
normal distribution 正态分布、高斯分布
normal equations 标准方程式
normal incidence 正入射
normal line 法线
normal matrix 正规矩阵
normal population 正态总体
normal probability curve 正态概率曲线
normal probability density function 正态概率密度函数
normal stress 正应力
normal vector 法向量
normalised data 数据库规范化
normalized 标准化的
norm 范数
notational expression 记号表达式
nought 零
nozzle-flapper mechanism 喷嘴挡板机构

null 零值的
number of oscillations 振荡次数
number 数
numerator 分子
numerical characteristics of random variables 随机变量的数字特征
numerical comparator 数值比较器
nyquist plot 奈奎斯特曲线
nyquist stability criterion 奈奎斯特稳定准则
object-oriented programming 面向对象编程
oblique 斜杠
observability 能观性
observable canonical form 可观标准型
observable decomposition 能观测性分解
octal 八进制的
octant 卦限
octave band 倍频带
octave 八度
odd function 奇函数
odd line 奇线
off-diagonal element 非对角元
off-line control 离线控制
off-resonant 偏离共振
offshore 近海的
off-the-shelf 现成的
Ohm's law 欧姆定律
omnidirectional 全向
on board 在船(火车、飞机、汽车)上
one-dimension probability distribution 一维概率分布
one-sided confidence interval 单侧置信区间
one-sided hypothesis 单侧假设,单边假设
one-sided limit 单边极限
one-sided test 单边检验,单侧检验

on-line control 在线控制
opaque 不透明的
open brace|open curly 左花括号
open bracket 左方括号
open interval 开区间
open parenthesis 左圆括号
open-loop control 开环控制
open-loop transfer function 开环传递函数
operational properties 操作性能
operational range 量程
operational rules of events 事件运算律
operator notation 算子记号
operator 运算符
optical 光学
optimal control 最优控制
optimization problems 最优化问题
ordinary differential equation 常微分方程
ordinary differential 常微分
ordnance 军械
orientation 方向
orifice 孔
original system 原系统
origin 原点
orthogonal 正交的
orthonormal basis 标准正交基
oscillating electric field 振荡电场
oscillation 振荡
out of phase 反相
outcome 结果
outer product 外积
outline 轮廓
out-of-balance 不平衡的
out-of-plane 平面外
output controllability 输出能控
over damp system 过阻尼系统
overhead 离地面的

overlaid 覆盖
overlap-save method 重叠保留法
overlay 在…上铺或盖
overshadow 使(某物)被遮暗
overtones 泛频
padded 填补
pad 垫
pairwise independent events 两两独立事件
pairwise independent 两两独立
palette 调色板
panel 面,板
parabola 抛物线
parabolic cylinder 抛物柱面
parabolic function 抛物线函数
paraboloid 抛物面
paradigm 范例
parallel connection 并联
parallel decomposition 并联分解
parallel lines 并行线
parallel resonance 并联谐振
parallelepiped 平行六面体
parallel 双线号
parameter estimation 参数估计
parameter identification 参数辨识
parameter of distribution 分布参数
parameter 参数
parametric hypothesis 参数假设
parametric model 参数模型
parametric statistical inference 参数统计推断
parametrization 参数化
parentheses 括号
partial derivative 偏导数
partial differential equation 偏微分方程
partial fractions 部分分式
partial integration 部分积分
particle velocity 粒子速度

particle 质点
partitioned 分割的
partition 分割
passive sensor 无源传感器
patch 小片
pattern search 模型搜索
payload 有效载荷
peak sound pressure 峰值声压
peak time 峰值时间
per mill 千分之
per cent 百分之
perceptron 感知器(模拟人类视神经控制系统的图形识别机)
perimeter 周长
periodic function 周期函数
periodogram 周期图
period 句号,点
period 周期
periphery 外围
permeability 可渗透的
permittivity 介电常数,电容率
permutation 序列,排列
perpendicular lines 垂直线
perpendicular to 垂直于
perpendicular 垂直的
personal digital entertainment products 个人数码娱乐产品
perspective 透视
perspex 塑胶
perturbation 摄动
pseudo plastic 伪塑性
phase current 相电流
phase margin 相角裕度
phase plane method 相平面法
phase shifter 移相器
phase shift 相移
phase to-neutral voltage 相电压
phase trajectory 相轨迹

phase-frequency characteristics 相频特性

phase-lag network 相位滞后网络

phasor 相位复(数)矢量

phenolic 酚的,石碳酸的

phonogram 表音符号

photoconductive effect 光导效应

physical insight 物理观点

physical principle 物理原理

piecewise defined function 分段定义函数

piezoceramic 压电陶瓷

piezoelectric & electrostrictive materials 压电和电致伸缩材料

piezoelectric effect 压电效应

piezoelectric strain gauges 压电应变计

piezoelectrics 压电

pillar 柱

pilot valve 操纵阀

pink noise 粉红噪声

pinning point 钉扎点

pinning 销连接

pipe clamp 管夹

pipes 管线

piston 活塞

pitching 俯仰

pitch 音调

pivoted 装在枢轴上的

pi 圆周率

plane force 平面力

plane wave 平面波

plane 平面

plastically deforming area 塑性变形区

plastic 塑性

platform 台,平台

played out 失去作用

plug-in 插件程序

plus or minus 正负

plus 加,正

ply (夹板的)层片

PN junction PN 结

pneumatic actuator 气动作动器

pneumatic system 气动系统

pneumatically 由空气作用

pneumatic 由压缩空气操作(推动)的

point estimate/point estimation 点估计

point estimator 点估计量

point force 点力

point in sample space 样本空间的点

point sources 点源

pointer 星指针

point-slope form 点斜式

poisson distribution 泊松分布

Poisson strain 泊松应变

Poisson theorem 泊松定理

Poisson's ratio 泊松比

polar axis 极轴

polar coordinates 极坐标

polar equation 极方程式

polarization 极化

polarized area 极化区域

pole point 极点

pole 极点

polyamide 聚酰胺

polygonization 多边化,多角形化

polymers 高分子材料

polymer 聚合物(体)

polynomial distribution 多项分布

polynomial 多项式

polyurethane 聚亚安酯

polyvinylidene fluoride 聚偏二氟乙烯

polyvinylidene 聚偏二乙烯的

populated 粒子数增加的

population central moment 总体中心距

population correlation coefficient 总体相关系数

population distribution 总体分布

population covariance 总体协方差
population mean 总体均值
population moment of order k 总体 k 阶矩
population moment 总体矩
population parameter 总体参数
population 总体
porous 能穿透的
ports-settings 端口设定
port 左舷
position vector 位置向量
positive angle 正角
positive definiteness 正定性
positive-going 正向的
posterior probability 后验概率
posteriori 后验
postprocessing 后加工
potential energy 势能
potential 电势
potentiometer 电位计
pound 井号
power amplifier 功率放大器
power factor 功率因数
power function 幂函数
power sensing device 功率传感器
power series 幂级数
power train 传动系
power 电源,功率
preamplifier (无线电)前置放大器
precision 精度
predictive control 预测控制
preestablished limit 预定的极限
pressure 压力
prestressed 预应力的(混凝土)
principal axes 主轴
principal stress 主应力
principle of duality 对偶性
prior probability 先验概率
priori 先验的

probability curve 概率曲线
probability distribution function 概率分布函数
probability density function 概率密度函数
probability density 概率密度
probability distribution 概率分布
processing 加工,处理
product theorem 乘法定理
product 积
profile 轮廓
programme 计划
proof mass 检测质量
proof-plane 验电板
propeller 螺旋桨
properties of state-transition matrix 状态转移矩阵的性质
properties of the actuator 作动器特性
prop-fan-engines 桨扇发动机
proportional controller 比例器
propulsion 推进
prototypical 原型的
prototyping 原型机制造
proximity sensors 临近传感器
pseudo-elastic approximation 拟弹性近似
pseudo-inverse 伪逆
pull-down menu 下拉选项屏(单)
pulsating sound source 脉动声源
pulse circuit 脉冲电路
punctuation 标点法
pure imaginary number 纯虚数
pure-tone 纯音
QR decomposition QR 分解
quadrant 四分之一圆
quadrant 象限
quadratically integrable function 二次判别函数

quadrature 正交
quadrilateral 四边(形)的
quadruple 四倍的,四重的
quadrupoles 四极子
qualitative control 定性控制
qualitative 定性的
quality of a sensor 传感器品质
quantifying 定量法
quasi-ideal 拟理想
quasistatic 准静态
question mark 问号
quiescent 静止
quotient law of limit 极限的商定律
quotient rule 商定律
quotient 商
rack 行李架
radiation impedance 辐射阻抗
radiation mode 辐射模态
radiation 辐射
radioactivity 放射性
radius of convergence 收敛半径
radius of curvature 曲率半径
rad 弧度
random event 随机事件
random noise 随机噪声
random occurrence 随机事件
random phenomenon 随机现象
random sample 随机样本
random sampling distribution 随机抽样分布
random sampling 随机抽样
random variables and distributions 随机变量及其分布
random variables 随机变量
range of a function 函数的值域
range 量程
rapidity 快速性
rate of change 变化率

rational function 有理函数
rational number 有理数
rationalizing substitution 有理代换法
Rayleigh's integral 瑞利积分
Rayleigh 瑞利
RC bridge type(venturi)oscillator RC 桥式(文氏)振荡器
react off 对抗
react on 对…起作用,对…有影响
reactance 电抗
reactive field 反应场
reactive power 无功功率
real number 实数
real symmetric matrix 实对称矩阵
real-time control 实时控制
receiver 听筒
receptance 接受率
reciprocating 往复的
reciprocity law 互易律
reciprocity principle 互易原理
reciprocity 相互作用
rectangle 矩形
rectangular coordinate system 直角坐标系
rectangular coordinates 直角坐标
rectification 整流
rectifier 整流器
recursion 递归
recursive least square(RLS)递归最小二乘方
recursive 递归的
redetermined size 预定尺寸
reference leak 标准漏孔
reference node 参考结点
reflected wave 反射波
reflection factor 反射因子
reflection/transmission coefficient of sound pressure 声压的反射系数与透射系数

reflection 反射
refraction 折射
regulator 稳压
reinforced concrete 钢筋混凝土
rejection region 拒绝域/否定区域
relative error 相对误差
relative maximum and minimum values 相对极大值与极小值
relative stability 相对稳定
relative stable 相对稳定
relay characteristics 继电特性
remanence 剩磁
repeatability 重复性
repetitive computation 重复(迭代)计算
reshape 改变…的形状
residual method 留数法
resilience 弹回
resin 树脂,松香
resistance change 阻抗变化
resistance temperature device(RTD) 电阻温度器件
resistance 电阻
resistive 阻抗
resolution 分辨率
resonance term 共振项
resonance 共振
resonant frequency 谐振频率
resonant muffler 共振式消声器
resonant peak value 谐振峰值
response time 响应时间
reticulated 网状的
revenue function 收入函数
reverberation radius 混响半径
reverberation room 混响室
reverberation time 混响时间
Reynolds number 雷诺数
rheological 流变学的

rheopectic substances 流凝性物质
ride quality control 乘坐品质控制
rider bar 游码标尺
riemann sum 黎曼和
Riemannian geometry 黎曼几何
right angle 直角
right-hand derivative 右导数
right-hand limit 右极限
rigid body 刚体
rise time 上升时间
robotics 机器人技术
robust control 鲁棒控制
roller 滚压机
rolling motion 滚动
root locus method 根轨迹法
root 根
rotational inertia 转动惯量
rotorcraft 旋翼飞机
rotor 转子
rounding error 舍入误差
Routh's stability criterion 劳斯稳定判据
rubber (合成)橡胶
rubber hardness tester 橡胶硬度测试器
rudder 舵
saddle point 鞍点
safety factor 安全系数
same distribution 同分布
sample average/sample mean 样本均值
sample central moment 样本中心矩
sample moment of order k 样本k阶矩
sample size 样本大小,样本容量
sample space 样本空间
sample standard deviation 样本标准差
sample value 样本值
sample variance 样本方差
samples and sampling distributions 样本及抽样分布

sample 样本
sampling control system 采样控制系统
sampling distribution 抽样分布, 样本分布
sampling statistics 样本统计量
sandwich 把…夹在…之间
saturation characteristic 饱和特性
saturation distortion 饱和失真
scalability 可量测性, 可伸缩性
scalar quantification 标量量化
scalar 标量
scaling 缩放比例
scanner 检测装置, 扫描设备, 扫描器
scatter 散射
scenario 方案
schematic 纲要的, 图表的
scheme 计划, 方案
script 脚本
scroll （电脑屏幕上）从上到下移动（资料等）
seaway 海上航道
secant line 割线
second derivative test 二阶导数试验法
second derivative 二阶导数
second partial derivative 二阶偏导数
secondary 次级的
second 秒
section 截面
seismic 地震的, 由地震引起的
self-radiation impedance 自辐射阻抗
self-adjoint 自伴的
self-calibration 自校准
self-contained 配套齐全的
self-diagnostics 自诊断
self-inductance 自感
self-radiation impedance 自辐射阻抗
semicircle 半圆
semicolon 分号
semiconductor integrated circuit 半导体集成电路
semiconductor transistor 半导体三极管
semiconductor 半导体
semi-positive definite 半正定
sensitive element 灵敏元件
sensitivity 灵敏度
sensor output 传感器输出
sensor 传感器
separation point 分离点
sequence control 顺序控制
sequential circuit 时序电路
series connection 串联
series decomposition 串联分解
series resonance 串联谐振
series 级数
servo system 伺服系统
settling time 调整时间
set 集合
shaker 激振机
Shannon's sampling theorem 香农采样定理
shape memory alloys 形状记忆合金
shaping filter 整形滤波器
shear connector 剪力接合器
shear force 剪切力
shear locking 剪力自锁
shear modulus 剪切模量
shear strain 剪切应变
shear stress 剪应力
shear velocity 剪切速度
sheet 薄板, 薄片
shell 壳
shift register 移位寄存器
short-haul 短程的
shunt 把（铁路货车等）调到另一轨道上
sideband 边（频）带

— 69 —

sifting 过筛
sign convention 符号规约
signal conditioning 信号调理
signal flow diagram 信号流图
signal source 信号发生器
signal 信号
significant level 显著水平
silhouette 轮廓
silicon strain gauges 硅应变计
silicone（聚）硅酮
silver 银
similar matrix 相似矩阵
simple frequency 简正频率
simple harmonic motion 简谐运动
simple random sample 简单随机样本
simple random sampling 简单随机抽样
simple support 简支
simply supported beam 简支梁
simulated annealing 模拟退火
simulate 仿真
simultaneous system 同步系统
since 因为
sine function 正弦函数
sine wave oscillator 正弦波振荡器
single channel control 单通道控制
single crystals 单晶体
single-input/single-output（SISO）单输入单输出
single quotation marks 单引号
singular value decomposition（SVD）奇异值分解
singular value 奇异值
singularity/singular point 奇点
singularity 奇点
sinusoidal signal 正弦信号
sinusoidal 正弦曲线的
size 尺寸
sketch 草图

skew hermite matrix 反 hermite 矩阵
skew plate 斜板
skew 歪
skyhook 架空吊运车
slab 厚板
slant asymptote 斜渐近线
slant 斜渐近线
slash-slash 双斜线
slash 斜线
sleeper pitch 轨枕间距
slide 滑动
sliding motion 滑动
slope function 斜坡函数
slope-intercept equation of a line 直线的斜截式
slope 斜率
smart materials 智能材料
smart system 智能系统
smear 涂抹
Smith normal form Smith 标准型
smooth curve 平滑曲线
smooth surface 平滑曲面
smoothen 使平滑
Snell acoustic reflection and refraction law 斯奈尔声波反射与折射定律
sodium potassium niobate（SPN）铌酸钠
soft boundary 软边界
software 软件
solenoid 螺线管
solid of revolution 旋转体
solution of homogeneous state equation 齐次方程的求解
solution of nonhomogeneous state equation 非齐次方程的求解
sonar 声呐装置,声呐系统
sound absorption coefficient 吸声系数
sound absorption 声吸收
sound energy density 声能量密度

sound energy 声能量
sound exposure level 声曝级
sound field 声场
sound holography 声全息
sound insulation test 隔声量测试
sound insulation 隔声量
sound intensity level 声强级
sound intensity method 声强法
sound intensity 声强
sound level meter 声级计
sound power absorption coefficient 声功率吸收系数
sound power calculation 声功率计算
sound power level 声功率级
sound power value 声功率值
sound power 声功率
sound pressure level 声压级
sound pressure 声压
sound 声
space 空间
span 跨度
spatial aliasing 空间混叠
spatio-temporal 时空的
special form of Chebyshev theorem 切比雪夫定理的特殊形式
special-purpose computer 专用计算机
specific heat 比热
spectral density 谱密度
spectral factorization 谱因子分解
spectral properties 谱性质
spectra 范围,光谱
spectrum 范围
spectrum 谱
speed of sound 声速
speed resonance frequency 速度共振频率
sphere 球体
spherical coordinate 球面坐标

spherical sound waves 球形声波
spherically 球状地
spike 尖状物
spin-off 副产品
spiral 螺旋形的
spool 管,筒,线轴
spring constant 劲度系数
springback angle 回弹角
spur gear 正齿轮
square brackets 方括号
square law error 平方律误差
square law 二乘法
square root 平方根
squared 方格的
square 正方形
squeeze theorem 夹挤定理
squeezed state 压缩态
stability analysis 稳定性分析
stability 稳定性
stable operating point 稳定工作点
stable state 稳态
stable 稳定
stack 堆
stain (使)染色
stamping 冲压
stand wave 驻波
stand-alone 单机
standard deviation 标准差
standard error 标准误差
standard normal curve 标准正态曲线
standard normal distribution 标准正态分布
standing wave field 驻波场
standing wave ratio 驻波比
standwave 驻波
starboard 右舷
start state 起始状态
star 星号

state controllability 状态能控
state feedback 状态反馈
state observer 状态观测器
state space 状态空间
state variable 状态变量
state vector 状态向量
state-space equation 状态空间表达式
state-transition matrix 状态转移矩阵
state 状态
static acceleration error coefficient 静态加速度误差系数
static characteristics 静态特性
static deflection 静变位,静荷载挠度
static model 静态模型
static operating point 静态工作点
static position error coefficient 静态位置误差系数
static state 静态
static velocity error constant 静态速度误差系数
statistical analysis 统计分析
statistical distribution 统计分布
statistical ensemble 统计总体
statistical hypothesis testing 统计假设检验
statistical hypothesis 统计假设
statistical inference 统计推断
statistical significance 统计显著性
statistic 统计量
steady state solution 稳态解
steady-state error 稳态误差
steady-state response 稳态响应
stealth 秘密行动
step function 阶跃函数
stern tube sealing oil pump 艉轴管轴封泵
stimulus 激励信号
stochastic sampling 随机抽样

stochastic signal 随机信号
stochastic 随机的
stove-pipe 烟囱管
strain sensor 应变计
strain 应变
stress 应力
strictly decreasing 严格递减
strictly increasing 严格递增
strip theory 薄片理论
strouhal number 斯特劳哈尔数
structure-borne noise 结构噪声
submarine 潜艇
submerged 在水中的
subroutine 子程序
subscript 下标的
substrate 基底,基片
subtraction 减
subtractor 减法器
sum 和
supercavitating flow 超空蚀水流
superconducting phenomenon 超导现象
superposition principle 叠加原理
superscript 上标
suppress 压制
supremum 最小上界
surface integral 面积分
surface of revolution 旋转曲面
surface transport 水陆运输
surface 曲面
susceptance 电纳
suspension stiffness 悬架刚度
swept meshing 扫掠网格
swung dash 代字号
symmetrical three-phase power supply 对称三相电源
synchronize (使)同步,(使)同速进行
synchronous optical network 同步光纤网

synchronous RS flip-flop 同步 RS 触发器

synchronous transfer mode 同步传输模式

system identification 系统辨识

System International of Units 国际单位制,简称 SI

system 系统

T flip-flop T 触发器

tabular 扁平的

tacho 转速计

tactile 触觉的

tailored 剪裁讲究的

tails of moving targets 运动目标移动痕迹

tangent function 正切函数

tangent line 切线

tangent plane 切平面

tangent vector 切向量

tan 黄褐色

Taylor expansion 泰勒展开

tear strength 抗扯强度

tectorial 顶盖的,覆膜的

temperature sensor 温度传感器

temperature 温度

temporary 暂时的

tensioning 张力调整

tension 张力

terbium 铽(符号 Tb)

Terfenol-D 稀土超磁致伸缩材料

test of goodness of fit 拟合优度检验

test of significance 显著性检验

test rig instrument 试验台仪表

test statistics 检验统计量

tetrahedron 四面体

the form of sensor output 传感器输出形式

the integral of 的积分

the inverse distance law 反距离法则

theorem 定理

thermal convective current 热对流

thermal 热

thermocouples 热电偶

thin-walled 薄壁的

thixotropic substances 触变性物质

threaded bolt 螺栓

three phase four wire system 三相四线制

three-phase A.C. circuit 三相电路

threshold of audibility 可听阈

throttle 控制油、气流的阀门

tie-rod 长螺栓

tilde 波浪符

tiling 盖瓦

timbre 音色

time constant 时间常数

time domain method 时域法

time history 随时间的变化

time-averaged 按时间平均的

time-dependent 时间相依

time-domain analysis 时域分析

time-harmonic wave 时谐波

time 时间

titanium 钛

Toeplitz matrix 托普利茨矩阵

toggle angle 肘节角

toolkit 工具包

tools 工具

tooth profile 齿廓,齿形

topology 拓扑

torque 扭矩

torsional constant 扭力常数

torsional strain 扭转应变

torsional 扭力的

total differential 全微分

total reflection(TIR) 全反射

track 轨道
trajectory 弹道
transducer 传感器
transfer function 传递函数
transfer mobility 传递迁移率
transformer 变压器
transient intermodulation distortion 瞬态互调失真
transient response analysis 暂态响应分析
transient response function 瞬态响应函数
transient solution 瞬态解
transient state 暂态
transistor 晶体管
translational 平动,平移
translucent 半透明的
transmissibility 可传送性
transmission cross section 透射截面
transmission of sound waves 声波的透射
transmitted beam 发射束
transpose 转置阵
transversal waves 横波
transversal 截线
transverse shear 横切力
transversely 横着地
transverse 横向的
trapezoidal 梯形的
trapezoid 梯形
triangle decomposition 三角分解
triangle 三角形
triangularly 成三角形地
triaxial 三轴的
trigonometric function 三角函数
trigonometric integrals 三角积分
trigonometric substitutions 三角代换法
trinomial distribution 三项分布

tripe integrals 三重积分
trough 低谷
truck 货车
truncate 截断的
truncation 切断
truss element 桁架杆元
tuned mass damper 调频质量阻尼器
turbo shaft engine 涡轮轴发动机
turbocharged diesel engine 涡轮增压柴油机
turbulent boundary layer(TBL) 紊流边界层
turbulent fluid 紊流
twist 扭
two types of errors 两类错误
two-dimensional continuous random variable 二维连续型随机变量
two-dimensional discrete random variable 二维离散型随机变量
two-dimensional normal distribution 二维正态分布
two-dimensional normal probability curve 二维正态概率曲线
two-dimensional normal probability density 二维正态概率密度
two-dimensional random variable 二维随机变量
two-point distribution 两点分布
two-position/on-off control 双位置、开关控制
two's complement 二进制补码
two-sided hypothesis 双侧假设
two-sided testing 双侧检验
two-sided 有两边的
U-distribution U 分布
U-estimator U 估计
ultrasonic wave 超声波
u-matrix 酉矩阵

unbiased estimate/unbiased estimation 无偏估计
uncorrelated 不相关的
undamped system 无阻尼系统
under damp system 欠阻尼系统
underlapped valve 负遮盖阀
uniform compression 均匀压缩
uniform distribution 均匀分布
unimodal 单峰的
uninsulated 未绝缘的
union events 并事件
union of 并, 合集
unit variance 单位方差
unitary low triangular 单位下三角矩阵
unitary matrix 酉矩阵
unitary space 酉空间
units 单位
unstable 不稳定
upper triangular matrix 上三角矩阵
urethane 聚氨酯
U-test U 检验
value of function 函数值
value range 值域
vapor 水汽
variable 变量
variance of random variables 随机变量的方差
variance ratio 方差比
variance 变异, 方差
varies as 与…成比例
vector quantization 矢量量化
vector space 向量空间
vector 向量
vehicle 车辆
vehicular 车的
velocit 速度
vertex 顶点
vertical asymptote 垂直渐近线

vertical bar 竖线
vertical virgule 竖线
vertically 垂直地
vessel 船
vibrant 振动的
vibration isolation 隔振
vibration pickup 拾振
vibratory 振动的
vibroacoustic environment 声振环境
vibrometer 振动计
viewport 视口
viscoelastic materials 黏弹性材料
viscoplastic flow 黏塑(性)流
viscous resistance 黏滞阻力
viscous 黏的
voice coil 音圈
voice-coil type actuator 音圈型调节器
voltage change 电压变化
voltage source 电压源
voltage-current curve 伏安曲线
voltage 电压
volume velocity 体积速度
volumetric 测定体积的
volume 体积
vortex sound theory 涡旋声理论
vorticity 涡度
wafer 薄片
wave node 波节
wave source 波源
wavefront 波前, 波阵面
waveguide 波导
wavelength 波长
waveloop 波腹
wavenumber filter 波数滤波器
wavenumber 波数
wax 蜡
wedge 楔形物

weight penalty 重量损失
weld 焊接,熔接
well balanced 均衡的
well posed problem 适定的问题
well-defined 定义明确的
well-developed 发育良好的
wheelhouse 舵手室
whistling 啸声
white box model 白箱模型
white noise 白噪声
wide-band amplifier 宽带放大器
wide-band filter 宽频带滤波器
widthwise 与宽同方向地,横向地
wildcard 通配符
wire current 线电流
wire harness 铠装线
wiring 接线
wishbone Y 字形的东西
woofer 低音用扩音器
work-energy theorem 功能原理
work 功
x-axis x 轴
x-coordinate x 坐标
x-intercept x 截距
yaw 偏航
yield stress 屈服应力
Z inverse transform Z 反变换
Z-distribution Z 分布
zero-order Bessel equation 零阶柱贝塞尔方程
zero drift problem 零点漂移问题
zero input response 零输入响应
zero phase shift 零相移
zero point 零点
zero state response 零状态响应
zero-crossing 零交点
zero-placement approach 零点配置方法

zeros of a polynomial 多项式的零点
Z-transformation Z 变换

附录 B 核对与检查计算公式和结果

有经验的人通常可以在较短时间内对计算公式和结果的正确与否做出判断，从而节省大量精力与时间。类似于编程过程中的语法测试、功能测试和性能测试，对计算公式和结果的判断也可分以上三步进行。

1. 语法测试。编程过程中的编译与连接操作，会对出现的语法错误进行提醒，这是最为基本的错误，在解析计算中其对应单位错误，这占据了可能出现错误的绝大部分，通常有以下规则：

● 平方根、立方根等内部项的单位必须是平方、立方等，例如 m^2、m^3 等。这并不代表内部项的每一项都是平方或立方等，只要其组合项满足以上要求即可，例如不考虑摩擦的自由落体的速度 $v = \sqrt{2gh}$，其中 g 的单位为 m/s^2，h 的单位为 m，从而 $2gh$ 的单位为 m^2/s^2，满足平方根下为平方的要求，开根号出来为 m/s，正好为速度的单位。

● 对数、指数以及三角函数等内部项必须满足无量纲的要求，不要求每一项均满足无量纲的要求，但是全部项组合而成的最终项必须满足无量纲的要求。

● 任意两个相加项必须具有相同的单位，例如质量 kg 不能与长度 m 相加等。

2. 功能测试。编程的目的是为实现某项功能，解析计算的目的也是给出合理的结果，同样有以下规则需要遵循：

● 确定结果的符号（+ 或 −）是否合理，例如计算某物的质量，若得到的结果为 −50 kg，则有必要检查计算过程了。

● 对于含有分式的方程，对分母进行检查是否有可能取值为零，即使检查结果表明其不可能为零，但是当分母减小，整个分式的值变大时也需做出是否合理的判断。

● 对式中的某项取值发生变化另一项也会随之变化时的合理性进行判断，例如分析固有频率的计算公式 $\omega = \sqrt{\dfrac{k}{m}}$，可以发现：

（1）刚度变大时，固有频率升高，合理；

（2）质量变大时，固有频率降低，也合理，等等。

● 对式中某项取特定值，例如 0，1 或无穷大时结果是否合理的情形进行判断，例如单自由度无阻尼受迫振动的解为 $x = A\sin(\omega_n t + \theta) + \dfrac{h}{\omega_n^2 - \omega^2}\sin(\omega t + \varphi)$，考虑 $\dfrac{h}{\omega_n^2 - \omega^2}$，则有当外部作用的频率 ω 取值等于系统的固有频率 ω_n 时，系统响应的振幅会趋于无穷大，这是合理的。

● 对于式中有两项相减的情形，考虑当二者取值相等时的结果是否合理，仍然以单自由度系统的受迫响应为例，当外部作用的频率 ω 等于系统的固有频率 ω_n 时，合理。

3. 性能测试。类似编程，快速地得到高精度的结果非常重要，这就有必要通过跟实验、其他人的结果或通过更为复杂过程计算得到的结果进行比较、判断。

附录 C 英语论文写作常用词汇

C.1 连接词汇

C.1.1 因果

therefore

Equation (1.3.2) is a second-order ordinary differential equation and therefore must have a solution which is specified in terms of two unknown constants or amplitudes of motion.

due to

The knowledge of the unit impulse response function allows us to derive expressions for the response of a mechanical system to any general excitation, including transients, and the response due to specified initial conditions using Duhamel's integral (Meirovitch, 1967; Newland, 1984).

result in

There are special cases that will result in a diagonal modal damping matrix, and therefore decouple the equations of motion.

since

The relationships between a wave and modal description of the response of a finite system is important to active vibration control since, as will be demonstrated in Chapter 6, effective control can often be achieved by actively changing the boundary conditions of the structure.

because

Such an approach will only be valid over a certain frequency region, however, because practical electronic integrators do not have the infinite gain at zero frequency ('d.c.') which an ideal integrator should possess.

hence

Hence, we see that the presence of bubbles around a ship may dramatically affect the sound propagation near the surface.

so that

We also assume that both the displacement and velocity at the two mounting points can be measured, so that we have as many outputs as state variables.

C.1.2 转折

rather than

In addition to these geometric requirements, this spacecraft would be sent in deep space (e.g. at the Lagrange point L2) rather than in low earth orbit, to ensure maximum sensitivity; this makes the weight issue particularly important.

however

However, its practical implementation is dependent upon development of realistic, time domain, wavenumber structural sensors that work over a broad frequency range.

instead of

As an alternative to the frequency-shaped cost functionals, loop shaping can be achieved by assuming that the plant noise has an appropriate power spectral density, instead of being a white noise.

although

Although most controller implementation is digital, current microprocessors are so fast that it is always more convenient, and sometimes wise, to perform a continuous design of the compensator and transform it into a digital controller as a second step, once a good continuous design has been achieved.

though

This does not mean that the control designer may ignore digital control theory, because even though the conversion from continuous to digital is greatly facilitated by software tools for computer aided control engineering, there are a number of fundamental issues that have to be considered with care; they will be briefly mentioned below.

unless

We are therefore certainly not closer to a solution unless we introduce some additional simplifying assumptions.

nevertheless

Nevertheless both Snyder et al. (1993) and Johnson and Elliott (1993) have developed and successfully tested distributed PVDF sensors designed to observe these radiation modes for the active control of structurally radiated sound.

otherwise

If the curves are intersecting curves between the limits of integration and situations arise where the upper curve switches and becomes a lower curve, then the integral representing the area must be broken up into integrals over sections otherwise one obtains a summation of "signed" areas.

C.1.3 加强

moreover

Moreover, in some applications like vibroacoustics, the behaviour of the structure itself is highly coupled with the surrounding medium; this also requires a coupled modelling.

furthermore

Furthermore, it is minimum phase and will have a stable inverse (see Nelson and Elliott, 1992, Chapter 3).

not only, but also

The phase change that a signal undergoes as it travels around the feedback loop is now not only determined by the jet but also by the delay in the acoustic response of the participating resonator.

further

This plasticity requires a mechanism for molecular mobility, which in crystalline materials can arise from dislocation motion (discussed further in a later chapter.)

in particular

In particular the interaction with the third cell appears to result into a localized periodic source of sound.

far more

The stress term line integral is far more difficult to assess.

in addition

In addition, a small amount of damping is usually added to the shell system to move the poles (associated with the eigenvalues) off the real axis thus avoiding the problem of numerical instability.

C.1.4 举例

begin with

We shall begin with the simple mass-spring system.

start with

Thus, we must start with linear differential EOMs by linearizing nonlinear EOMs.

for example

For example, a system with Coulomb friction is governed by the type of damping characteristic shown in Figure 3.4.

for instance

For instance, an offshore oil platform (Figure 3.6) vibrates as if it were a SDOF system in the case of many circumstances because the support acts largely as a stiffness element and the platform acts like an inertia in the frequency range associated with typical waves.

e.g.

When other arbitrary forms of damping (e.g. structural) are encountered, then the equivalent viscous damping can be computed for harmonic inputs as described previously in Section 3.2, and then the Duncan-Collar formulation can be used to perform the modal transformation.

etc.

This system could represent a clothes washer or dryer with an imbalance in how the clothes are distributed, a lathe or rotating machine tool with a slight imbalance, or any other rotating system in which the center of rotation does not coincide with the center of mass(e.g. fan, disk drive, etc.).

C.1.5 承上启下

i.e.

This motivates the desire to design a control system which does not affect the form of these modes (i.e. does not change the eigenvectors of A) but provides independent control over the natural frequencies and damping of these modes(i.e. allows modification of the eigenvalues of A).

so called

By placing the sensor and actuator at the same point on the structure the effect of velocity feedback is the same as if a passive damper had been attached between that point on the structure and an inertial reference; the so-called skyhook damper(Kamopp et al., 1974).

in terms of

Rather than directly transforming the differential equations which describe a dynamic system into the Laplace domain, an alternative approach is to recast the time domain equations into a standard form; in terms of the internal state variables of the system.

this implies that

In practice therefore, this implies that the secondary force would have to be applied to the body of the vibrating machine via an inertial exciter of some kind(e.g. an electrodynamic exciter which is provided with a mass against which it can react).

apart from

In general the procedure is similar to that described in Section 9.4 apart from a number of important differences.

aforementioned

This effect can often work to advantage in curtailing the aforementioned control spillover problem.

with the exception of

Good attenuation is obtained over a frequency range of 0-650 Hz with four actuators with the exception of the peak near 330 Hz.

in the previous section

Although it is possible to derive the frequency response of the optimal feedforward controller using the methods outlined in the previous sections, the problem of designing a practical filter which implements this frequency response still remains.

in general

In general the attenuation of sound waves increases with frequency.

this exsplains why

This explains why we hear the lower frequencies of an airplane more and more accentuated as it flies from near the observation point (e.g. the airport) away to large distances (10 km).

in the presence of

In the presence of walls the viscous dissipation and thermal conduction will result into a significant attenuation of the waves over quite short distances.

as a general rule

As a general rule, at low amplitudes the viscous dissipation is dominant in woodwind instruments at the fundamental (lowest) playing frequency.

it should be noted

It should be noted that although the theory is derived in terms of piezoelectric material, it is generally applicable to all distributed strain-inducing actuators and sensors.

it is interesting to see that

It is interesting to see that shear stresses are predicted, whose origin is based on the intramolecular (elastic) interaction between the beads

in many cases

In many cases, when the acoustic wave length is small compared to the temperature gradient length (distance over which a significant temperature variation occurs) we can still use the wave equation (2.17a).

as a rule of thumb

As a rule of thumb, the observer poles should be 2 to 6 times faster than the regulator poles.

state-of-the art

In this paper, we present a survey on the state of the art knowledge on this topic, which is incomplete, and indicate some new trends for further research.

C.2 常用动词及动词短语

satisfy

The real part of the exponent in equation (8.3.6) must always be negative in order to satisfy the Sommerfeld radiation condition; see Junger and Feit, 1986, Ch. 5.

correspond to

This corresponds to the complex velocity distribution associated with a simply supported plate vibrating in its (m, n) mode as discussed in Chapter 2.

allow for

Before we do this, it is necessary to

extend the previously described cylinder equations of Sections 2.12 and 2.14 to allow for coupled interior and radiated acoustic fields, various forcing functions and the effects of finiteness of the cylinder.

similar to

James (1982) has used a procedure similar to that described above to find the pressure and response of a cylinder excited by an exterior monopole.

formulate

In this section we formulate the equations of motion in matrix form.

depend on

Darlington has shown that the relative heights of the two peaks in the frequency response of the system, on either side of the reference frequency, depend on the phase error of the secondary path model, and suggests that this asymmetry could be exploited as a diagnostic tool to detect such phase errors.

derive

It should be noted that although the theory is derived in terms of piezoelectric material, it is generally applicable to all distributed strain-inducing actuators and sensors.

construct

The piezoelectric elements were of length 38.1 mm, 15.8 mm width and 0.2 mm thickness and were constructed from a ceramic material, G1195 with properties specified in Table 5.1. approximate.

reduce

Thus, the dynamic problem is reduced to an "equivalent" problem of statics.

abatement

This article brings together the noise data of part of our inland ships, introduces the permissible standard of ship noise abroad, and tells of the various methods of controlling ship noise. All the data collected here may be for reference use in the design of ship noise abatement.

attenuate

Solution (b) consists of two complex solutions at low frequencies which when combined together form an attenuated standing wave (or near field).

augment

The use of a disturbance model to augment the vector of state variables is described by Johnson (1976), and had been previously used in the active control of a torsional system, for example, by Burdess and Metcalfe (1985).

contribute to

Transient terms eventually decay and disappear and do not contribute to the solution after about 5 time constants.

diminish

Since both the roots are negative, the motion diminishes with increasing time and is aperiodic.

eliminate

This projection would effectively eliminate the constraint but retain the active forces as desired.

outperform

Simulation and experiments have been done for images contaminated by various noise level, and the results show that the proposed algorithm outperform the median filter and adaptive-neighborhood filter.

implement

ANC that uses adaptive signal processing

implemented on a low-cost, high-performance DSP is an emerging new technology.

implementation

These devices have enabled the low-cost implementation of powerful adaptive ANC algorithms and encouraged the widespread development of ANC systems.

suffice for

The experimental results that accuracy of this method suffice for pactice.

introduce

Air bubbles are also introduced in sea water near the surface by surface waves.

involve

The dynamics of bubbles involving oscillations (see chapter 4 and chapter 6) appear to induce spectacular dispersion effects[42], which we have ignored here.

expect

We have seen in section 2.3 that the speed of sound in the atmosphere is expected to vary considerably as a result of temperature gradients.

prove

This proves that the measurement of the acoustic field outside the source region is not sufficient to determine the source uniquely[52].

refer to

The notion of "analogy" refers here to the idea of representing a complex fluid mechanical process that acts as an acoustic source by an acoustically equivalent source term.

base on

Such estimation procedure is based on the physical interpretation of the source term.

stress

It should again be stressed here that in ASAC, although the control action is applied directly to the structure, the cost function is derived from the far-field radiated pressure (or far-field radiated pressure-related variables).

verify

The values of the matrices Mand are entered (READ) and verified (WRITE) in Par. #I.

denote

In mechanics, impedance denotes originally the ratio between a force amplitude and a velocity amplitude.

evaluate

They are evaluated by applying the initial conditions to the general solution in Eq(2-47).

use

The method is generally used for vibration measurement.

apply

They are evaluated by applying the initial conditions to the general solution, since it is the entire solution that must satisfy the initial conditions.

utilize

Although the concept of equivalent quantities may not be fully utilized in this chapter, they are introduced early in the text because (1) the one-degree-of-freedom system is basic in vibration, and (2) the concept of equivalent or generalized quantities is essential for more advanced studies in later chapters.

whereby

For $\beta \neq 0$, there can still be a resonance-type behavior whereby the amplitude of the oscillations become large for some specific value of the forcing frequency λ.

withstand

For instance, if buildings were designed

only to support their own weight and not to withstand dynamic excitations, they would fail catastrophically if subjected to base excitations during earthquakes.

C.3 常用副词

predominantly

The purpose of presenting the resonance information in this form is that it illustrates the interesting result that for the $s = 1$ (flexural) branch which corresponds to predominantly radial motion, the resonance frequency initially falls and then increases with n.

accordingly

Accordingly, if the QWSIS is used as an actuator, it is equivalent to a uniform pressure actuator (Fig. 4.3.c).

macroscopically

The effective properties of macroscopically nonuniform ferromagnetic composites: theory and numerical experiment.

especially

Most approaches to problems of this type have involved the application of feedback control, since disturbances causing the vibration (especially in the case of road vehicles) are generally random.

drastically

This results in a drastically different sound radiation pattern.

intuitively

This characteristic is also intuitively obvious when one examines the spatial phase distribution of the $n = 2,4,6,\ldots$ or antisymmetric modes.

considerably

In structures, however, the wavespeed of compressional disturbances is generally considerably larger than that for acoustic waves in air, and the high frequency components of flexural waves can also propagate very rapidly.

mathematically

Mathematically it is important to note that an impedance boundary condition is of "mixed type".

extensively

The "singing flame" has already been discussed extensively by Rayleigh.

typically

Typically, the multiple scale method is applicable to problems with on the one hand a certain global quantity (energy, power) which is conserved or almost conserved and controls the amplitude, and on the other hand two rapidly interacting quantities (kinetic and potential energy) controlling the phase.

essentially

The control system (which was essentially feedforward in nature) operated in the frequency domain, using a discrete Fourier transform of the measured accelerations, essentially to establish the complex matrix relating the secondary force inputs to the secondary acceleration outputs and the vector of complex accelerations produced by the primary excitation.

substantially

Above this frequency, however, the effectiveness of the passive isolation increases substantially.

Obviously

Obviously the inclusion and characteristics of damping are very important to active control methods since it represents a process by which the response of a system can also be reduced by passive means.

usually

Piezoelectric coefficients, usually written in a form with double subscripts, provide the relationship between electrical and mechanical quantities.

historically

This can be a tedious job if performed manually and although, these days, computers can be readily programmed to perform this task, a number of computational methods have historically been used to manually determine the signs of the real parts of all the roots of the characteristic equation.

traditionally

Traditionally, the angle used is between the propagation direction and the normal vector of the interface.

conventionally

A further benefit emphasised by Mc Donald et al. is the improvement in ride and handling qualities of a vehicle whose engine mounts can be made much stiffer than those used conventionally.

progressively

An approach to adaptive ANC performance analysis that involves a hierarchy of techniques, starting with an ideal simplified problem and progressively adding practical constraints and other complexities, was developed by Morgan[8].

C.4 讨 论

in the sense

This approach is identical to feedforward control in the sense discussed here.

obtain

Displacement and acceleration can be obtained from the electrical output of the velocity pickup by integration and differentiation.

derive

The derived units are formed from the base units according to the algebric relations linking the corresponding quantities.

simplify

Make the necessary assumptions to simplify the problem and deduce the equation of motion for the flapping motion of the blade.

develop

The emphasis is on problem formulation and interpretation, since the general theory was developed in the last chapter.

substitute

Once the forcing pressure field has been expanded into the appropriate functions (i.e. the fight hand side of equation (9.7.10)), the functions can be substituted into the fight hand side of the radial equation of the shell equations which in this case are the Donnell-Mushtari relations (see equation (2.12.1c)).

compared with

This is the characteristic of the frequency equation, which may be compared with Eq. (4-13).

in contrast

Note that a negative sign has been used in equation (1.2.2) in contrast to many texts dealing with vibration.

discuss

This problem will be discussed further in the next chapter.

study

For illustrative purposes we study two example forcing conditions.

extend

It is of interest to extend the scope of

our discussions to consider elastic motion in two dimensions, i. e. the vibrations of thin plates or panels.

as long as

Clearly as N becomes larger one obtains a better approximation to the global response of the structure as long as the sensors are appropriately distributed.

yield

Equating the sum of the appropriate elements of each of the matrices to zero yields four independent equations, two of which are identical.

generalization

This generalization of Howe's equation is indeed derived by Jenvey[96].

agree with

Rearrange the axes in figure 1-23(b) so the x-axis is to the right and the y-axis is vertical so that the axes agree with the axes representation in figure 1-23(a).

suppose

Suppose it is required to undo what has just been done.

assume

The integration is with respect to the variable t and it is assumed that the integrand f is both continuous and differentiable with respect to x.

gin order to

In order to use these tables one must sometimes make appropriate substitutions in order to convert an integral into the proper form as given in the tables.

lead to

Some notations use a different starting index which can lead to confusion at times.

take into account

Furthermore, energy methods only take into account active forces; idealized forces of constraint, which do no work, are ignored.

in detail

We will not consider the generation process herein detail, but only indicate the presence of the eigensolution for a distinct source far upstream.

focus on

We therefore focus our attention in this chapter on the one-dimensional approximation of duct acoustics.

imply

This implies that there is no transmission of sound through these walls.

alternative

As an alternative we now show the wave equation in characteristic form.

move to

Increasing damping also causes the frequency of maximum response to move to lower values.

reasonable

This assumption was not justified but it seems reasonable if the acoustic velocities in the flow are "small enough".

point of view

Under these conditions the system is escribefl as being flriven onresonance anti is a very important conflition from a control point of view.

preliminary

Charts are presented which enable preliminary design calculations to be undertaken graphically.

limit to

We will now further limit our discussion to the case of harmonic waves.

overall

This second mechanism of control, where the overall plate vibration amplitude is not significantly attenuated or sometimes increased while supersonic wavenumber components, and the associated sound radiation are reduced, we term control by modal restructuring (Fuller et al. ,1991).

non-trivial

In these cases, the conditions of stability to plane perturbations are non-trivial and are investigated below.

C.5 图 表

C.5.1 图表描述

diagrammatic

Figure 9.13 shows a diagrammatic representation of a plan view of the rear of the DC-9 test fuselage.

describe

Alternatively, the motion of a system can be analysed in terms of the structural waves which propagate within it, and the active control of structural waves is described in the second half of this chapter.

show

The adaptivity of the feedforward controller ensures that it is not "open loop", and a brief analysis is presented that shows how adaptive feedforward controllers can be represented as equivalent feedback systems.

demonstrate

The considerable promise shown by the application of active techniques to these problems is clearly demonstrated.

tell

This line tells one that as increases (price increases), then the number of sales x decreases.

propose

For gases which are imperfect, there are many other proposed equations of state.

see

Critical points must then be tested to see if they correspond to a local maximum, local minimum or neither, such as the point x_7 in figure 2-12.

note

Note that a negative sign has been used in equation (1.2.2) in contrast to many texts dealing with vibration.

illustrate

Although the above analysis is straightforward, it does illustrate the basic process by which elastic systems are generally analysed.

denote

It is convenient to adopt complex notation to denote the amplitude and phase of the various signals, and of the frequency responses of the mechanical paths at the reference frequency.

observe

Choice of the particular sensor configuration is dependent upon the system variable to the observed, and to some degree, the form of signal processing to be used.

indicate

Thus, the results indicate that in order to dominantly drive longitudinal motion with this configuration, longer, extended arrays of actuators are needed.

provide

If this stiffness is included in the

analysis (with applied voltage set to zero on that element) then it can be shown that the anti-symmetric actuator provides exactly twice the input moment.

find

Liang et al. use a Rayleigh-Ritz approach to find the normal mode response of a simply supported rectangular plate with embedded SMA fibres by including the recovery stress due to activation of the SMA fibres directly into the governing laminate equations (the ASET principle).

investigate

In their work they investigated the use of IIR fixed filters as plant models (Vipperman et al., 1993) and studied the effects of non-causality due to delays through the control hardware, as also discussed in Section 6.13 of Nelson and Elliott.

predict

Chapter 8, on nonlinear systems, explains certain common phenomena that cannot be predicted by linear theory.

data

The result of a computation cannot have any more significant numbers than that in the original data.

figure

The relative positions of the vectors are illustrated in the figure.

draw

A mnemonic device to aid in calculating the determinant of the coefficients is to append the first two columns of the coefficients to the end of the array and then draw diagonals through the coefficients.

C.5.2 变化趋势

	动词表达	用名词作同义替换表达
上升	grow(to)	a growth(in)
	rise(to)	a rise(in)
	increase(to)	an increase(in)
	go up(to)	a growth(in)
	climb(to)	an upward trend(in)
	boom	a boom in (a dramatic rise in)
下降	fall(to)	a decrease(in)
	decrease(to)	a decline(in)
	dip(to)	a drop(in)
	drop(to)	
	go down(to)	a slump in (a dramatic fall in)
	slump(to)	
	reduce(to)	a reduction(in)
徘徊	level out(at)	a leveling out(of)
	did not change	no change(in)
	remain stable(at)	
	remain steady(at)	
	stay constant(at)	
	maintain the same level	
波动	fluctuate(around)	a fluctuation
	peak(at)	reach a peak
	plateau(at)	reach a plateau
	stand at	

C.5.3 变化程度

形容词	副词
Dramatic	Dramatically
Sharp	Sharply
Huge	hugely
enormous	enormously
steep	steeply
substantial	substantially
considerable	considerably
significant	significantly
marked	markedly
moderate	moderately
slight	slightly
small	smally
minimal	minimally

C.5.4 变化速度

形容词	副词
rapid	rapidly
quick	quickly
swift	swiftly
sudden	suddenly
steady	steadily
gradual	gradually
slow	slowly
occasional	occasionally

参 考 文 献

[1] ARBEL A. Controllability measures and actuator placement in oscillatory systems[J]. International Journal of Control,1981,33(3):565 – 574.

[2] BADEL A,SEBALD G,GUYOMAR D, et al. Piezoelectric vibration control by synchronized switching on adaptive voltage sources:Towards wideband semi-active damping[J]. Journal of the Acoustical Society of America,2006,119(5):2815 – 2825.

[3] BERRY A,QIU X,HANSEN C H. Near-field sensing strategies for the active control of the sound radiated from a plate[J]. Journal of the Acoustical Society of America,1999,106(6):3394 – 3406.

[4] HAC A,RADKOWSKI S. Applicatioin of Kalman filter technique to measurement of stochastic vibration of distributed parameter systems [R]. Institute of Machine Design Fundamentals Research,1982 (13):113 – 137.

[5] HAC A,LIU L. Sensor and actuator location in motion control of flexible structures [J]. Journal of Sound and Vibration,1993,167(2):239 – 261.

[6] MIGUEZ-OLIVARES A A,RECUERO-LOPEZ M. Development of an active noise controller in the DSP start kit [R]. The first European DSP Education and Research Conference,1996.

[7] BERKHO F F A P, WESSELINK J M. Combined MIMO adaptive and decentralized controllers for broadband active noise and vibration control[J]. Mechanical Systems and Signal Processing,2011,25(5):1702 – 1714.

[8] MONTAZERI A,POSHTAN J. A computationally efficient adaptive IIR solution to active noise and vibration control systems[J]. IEEE Transactions on Automatic Control,2010,55 (11):2671 – 2676.

[9] ANDRÉ PREUMONT. Vibration Control of Active Structures, AnIntroduction [M]. 3rd. Springer,2002.

[10] YOUNG A J. Active control of vibration in stiffened structures[D]. Adelaide:The University of Adelaide South Australia,1995.

[11] BAEK K H,ELLIOTT S J. Natural algorithms for choosing source locations in active control systems[J]. Journal of Sound and Vibration,1995,186(2):245 – 267.

[12] VEL S S, BAILLARGEON B P. Active vibration suppression of smart structures using piezoelectric shear actuators[D]. Maine:The University of Maine,2002.

[13] BANKS H T,et al. A piezoelectric actuator model for active vibration and noise control in thin cylindrical shells [C].1992.

[14] BEIN T. Intelligent Materials for active noise reduction-overview & results[C]// Transport Research Arena,2006.

[15] BENJAMIN C K,GOLNARAGHI F. 自动控制系统[M]. 汪小帆,李翔,译. 北京:高等教育出版社,2004.

[16] BERKHOFF A P,WESSELINK J M. Combined MIMO adaptive and decentralized controllers for broadband active noise and vibration control[J]. Mechanical Systems and Signal Processing,2011,25:1702-1714.

[17] WIDROW B,SAMUEL D S. Adaptive Signal Processing[M]. 王永德,龙宪惠,译. 北京:机械工业出版社,2008.

[18] LI B,MOORE S. Noise control for fluid power systems[C]// Internoise-international Congress on Noise Control Engineering,2014.

[19] BIRMAN V. Active control of composite plates using piezoelectric stiffeners[J]. International Journal of Mechanical Sciences,1993,35(5):387-396.

[20] BRAGHIN F,CINQUEMANI S,RESTA F. A model of magnetostrictive actuators for active vibration control[J]. Sensors and Actuators A Physical,2011,165(2):342-350.

[21] BRENNAN A M C,MCGOWAN A M R. Piezoelectric power requirements for active vibration control[J]. Proceedings of SPIE-The International Society for Optical Engineering,1997,114(9):1542-1570.

[22] BRODY D. JOHNSON,CHRIS R F. Broadband control of plate radiation using a piezoelectric double-amplifier active-skin and structural acoustic sensing[J]. Acoustical Society of America,2000,107(2):876-894.

[23] USTUNDAG B,HARDT D E. On the free vibration behavior of cylindrical shell structures[D]. Massachusetts Institute of Technology,2011.

[24] GENTRY C A. Smart foam for applications in passive-active noise radiation control[J]. Acoustical Society of America,1997,101(4):1771-1778.

[25] MAURY C,GARDONIO P,ELLIOTT S J. A Wavenumber Approach to Modelling the Response of A Randomly Excited Panel,Part I:General Theory[J]. Journal of Sound and Vibration,2002,252 (1):83-113.

[26] MAURY C,GARDONIO P,ELLIOTT S J. A Wavenumber Approach to Modelling the Response of A Randomly Excited Panel,Part II:Application to Aircraft Panels Excited by a turbulent Boundary Layer[J]. Journal of Sound and Vibration,2002,252(1):115-139.

[27] PAULITSCH C,GARDONIO P,ELLIOTT S J,et al. Design of a lightweight,electrodynamic,inertial actuator with integrated velocity sensor for active vibration control of a thin lightly-damped panel[C]. ISMA Publications,2004.

[28] FULLER C R,ELLIOTT S J,NELSON P A. Active Control of Vibration[M]. Academic Press,1997.

[29] FULLER C R,ELLIOTT S J,NELSON P A. Active Control of Vibration[M]. Elsevier Ltd,1996.

[30] LAI C. Acoustic radiation from finite length cylindrical shells using boundary element method[C]. Australia:Fifth International Congress on Sound and Vibration,1997.

[31] YIN C, SUN H, AN F, et al. Active control of low-frequency sound radiation by cylindrical shell with piezoelectric stack force actuators[J]. Journal of Sound and Vibration, 2012, 331(11):2471-2484.

[32] CARESTA M, KESSISSOGLOU N K. Structural and acoustic responses of a fluid-loaded cylindrical hull with structural discontinuities[J]. Applied Acoustics, 2008, 70(7):954-963.

[33] BLOCKA C, ENGELHARDTA J, HENKELA F. Active control of vibrations in piping systems[C]. 20th International Conference on Structural Mechanics in Reactor Technology, 2009.

[34] BERT C W, KIM C D, BIRMAN V. Vibration of composite-material cylindrical shells with ring and/or stringer stiffeners[J]. Elsevier, 1993, 25(1-4):477-484.

[35] JORDAN C, STEVE D. Broadband active control of noise and vibration in a fluid-filled pipeline using an array of non-intrusive structural actuators[J]. INTER-NOISE and NOISE-CON Congress and Conference Proceedings, 2015, 250(3).

[36] CHEN G S, BRUNO R J, SALAMA, et al. Optimal placement of active/passing members in truss structures using simulated annealing[J]. AIAA Journal, 1991, 29(8):1327-1334.

[37] CHHABRA D, NARWAL K, SINGH P. Design and analysis of piezoelectric smart beam for active vibration control[J]. International Journal of Advancements in Research and Technology, 2012, 1:1-5.

[38] CHRISTOPHER C. Active Control of Vibration[M]. Academic Press, 1997.

[39] CHRISTOPHER E. RUCKMAN. Optimizing actuator locations in active noise control systems using subset selection[J]. Journal of Sound and Vibration, 1995, 186(3):395-406.

[40] HANSEN C H, SNYDER S D. Active Control of Noise and Vibration. Second Edition[M]. CRC Press, 2012.

[41] CUMMINGS A. Sound transmission through duct walls[J]. Journal of Sound & Vibration, 2001, 239(4):731-765.

[42] CYRIL M. HARRIS, ALLAN G. Piersol. Harris'Shock and Vibration Handbook[M]. The McGraw-Hill Companies, 2002.

[43] THOMAS D R, NELSON P A, ELLIOTT S J. Active control of the transmission of sound through a thin cylindrical shell, part1: the minimization of vibrational energy[J]. Journal of Sound and Vibration, 1993, 167(1), 91-111.

[44] DAVID A B, COLIN H H. Engineering Noise Control: Theory and Practice.[M]. Fourth Edition. CRC Press, 2009.

[45] DENG Q, JIANG W, TAN M, et al. Modeling of offshore pile driving noise using a semi-analytical variational formulation[J]. Applied Acoustics, 2016, 104:85-100.

[46] YANG D Q, YI G L, CHENG J P. Multidisciplinary design optimization of sound radiation from underwater double cylindrical shell structure[C]. Sydney: 11th World Congress on Structural and Multidisciplinary Optimisation, 2015.

[47] DROŻYNER P. Determining the limits of piping vibration[J]. Physical Review: a, 2011, 84

(5):285-297.

[48] EDWARD R. Scheimerman. Invitation to Dynamical Systems [M]. Dover Publications, 2012.

[49] PERINI E A. Active control in rotating machinery using magnetic actuators with linear matrix inequalities (LMI) approach [C]. Proceedings of the IMAC-XXVII, 2009.

[50] FARSHIDIANFAR A, OLIAZADEH P. Free vibration analysis of circular cylindrical shells: comparison of different shell theories[J]. International Journal of Mechanics and Applications, 2012, 2(5):74-80.

[51] GROSVELD F W. Numerical comparison of active acoustic and structural noise control in a stiffened double wall cylinder [J]. American Institute of Aeronautics and Astronautics, 1996.

[52] FLÜGGE W. Stresses in Shells [M]. Berlin: Springer-Verlag, 1960.

[53] ZHAO G, ALUJEVIC N, DEPRAETERE B, et al. Experimental study on active structural acoustic control of rotating machinery using rotating piezo-based inertial actuators[C]. Journal of Sound and Vibration, 2015, 348.

[54] GIBBS G P, CLARK R L, COX D E, et al. Radiation modal expansion: Application to active structural acoustic control [J]. Acoustical Society of America, 1999, 107(1)3:32-339.

[55] 博克斯,詹金斯,莱茵泽尔,等. 时间序列分析:预测与控制[M]. 王成璋,译. 北京:机械工业出版社,2011.

[56] GORMAN D G, REESE J M, ZHANG Y L. Vibration of a flexible pipe conveying viscous pulsating fluid flow[J]. Journal of Sound & Vibration, 2000, 230(230):379-392.

[57] OLSON H F. Electronic control of noise, vibration, and reverberation[J]. J. acoust. soc. am, 1956, 28(1):966-972.

[58] OLSON H F, MAY E G. Electronic sound absorber [P]. U. S. Patent US 2,983,790, filed: April 30, 1953, patented: May 9, 1961.

[59] HASAN KORUK, JASON T. Modal analysis of thin cylindrical shells with cardboard liners and estimation of loss factors [J], Mechanical Systems and Signal Processing, 2014, 45(2).

[60] HOLLAND J H. Adaptation in Natural and Artificial Systems[M]. Cambridge: MIT, 1975.

[61] DAI H L, JIANG H J. Forced vibration analysis for a FGPM cylindrical shell[J]. Shock and Vibration, 2012, 20(3):531-550.

[62] SHEN H J, WEN J H, Yu D L, et al. Control of sound and vibration of fluid-filled cylindrical shells via periodic design and active control[J]. Journal of Sound and Vibration, 2013, 332(18):4193-4209.

[63] IZUCHI H, NISHIGUCHI M, LEE G Y H. Fatigue life estimation of piping system for evaluation of acoustically induced vibration (AIV) [C] INTER-NOISE and NOISE-CON Congress and Conference Proceedings, 2014, 249(8).

[64] WACHEL J C, MOTON S J, KENNETH E. Atkins, piping vibration analysis [C]. Proceeding of the Nineteenth Turbomachinery Symposium, 1990.

[65] DEAN, J, GIBBS M R J, SCHREFL T, et al. Finite-element analysis on cantilever beams coated with magnetostrictive Material[J]. IEEE Transactions on Magnetics, 2006, 42(2):

283-288.

[66] JUANG J N, RODRIGUEZ G. Formualtions and Dynamcis and Control of Large structures actuator and sensor placements [J]. Proceedings of the second YPI&SU/AIAA Symposium on Dynamics and Control of Large Flexible Spacecraft, 1979, 247 – 262.

[67] MAILLARD J P, FULLER C R. Advanced time domain wave – number sensing for structural acoustic systems. part III. experiments on active broadband radiation control of a simply supported plate[J]. Journal of the Acoustical Society of America, 1995, 98(5): 2613 – 2621.

[68] MAILLARD J P, FULLER C R. Acitve control of sound radiation from cylinders with piezoelectric actuators and structural acoustic sensing [J]. Jounal of Sound and Vibration, 1997, 222(3): 363 – 388.

[69] CARNEAL J P, CHARETTE F, FULLER C R. Minimization of sound radiation from plates using adaptive tuned vibration absorbers[J]. Journal of Sound and Vibration, 2004, 270(4):: 781 – 792.

[70] CARNEAL J P, GIOVANARDI M, FULLER C R, et al. Re – active passive devices for control of noise transmission through a panel[J]. Journal of Sound and Vibration, 2008, 309(3 – 5): 495 – 506.

[71] CARNEAL J P, FULLER C R. An analytical and experimental investigation of active structural acoustic control of noise transmission through double panel systems[J]. Journal of Sound and Vibration, 2004, 272(3 – 5): 749 – 771.

[72] CHEN J C, BABCOCK C D. Nonlinear vibration of cylindrical Shells[J]. Aiaa Journal, 1975, 13(7): 868 – 876.

[73] PAN J, SNYDER S D, HANSEN C H, et al. Active control of far - field sound radiated by a rectangular panel-a general analysis[J]. The Journal of the Acoustical Society of America, 1992, 91(4): 2056 – 2066.

[74] DOYLE J C, FRANCIS B A, TANNENBAUM A R. Feedback Control Theory [M]. Macmillan Publishing, 1990.

[75] MOTTERSHEAD J E, RAM Y M. Inverse eigenvalue problems in vibration absorption: Passive modification and active control [J]. Mechanical Systems and Signal Processing, 2006, 20(1): 5 – 44.

[76] NAUMANN E C, SEWALL J L. An experimental and analytical vibration study of thin cylindrical shells with and without longitudinal stiffeners [R]. Washington D C: national aeronautics and space administration 9, 1968.

[77] JOVANOVIC M M, SIMONOVIC A M, ZORIC N D, et al. Experimental Investigation of spillover effect in system of active vibration control[J]. Fme Transactions, 2014, 42(4): 329 – 334.

[78] NIU J, SONG K, LIM C W. On active vibration isolation of floating raft system[J]. Journal of Sound and Vibration, 2005, 285(1 – 2): 391 – 406.

[79] SOHN J W, CHOI S B, KIM H S. Vibration control of smart hull structure with optimally

placed piezoelectric composite actuators[J]. International Journal of Mechanical Sciences, 2011,53(8):647-659.

[80] LIEW K M,HE X Q,KITIPORNCHAI S. Finite element method for the feedback control of FGM shells in the frequency domain via piezoelectric sensors and actuators[J]. Computer Methods in Applied Mechanics & Engineering,2004,193(3-5):257-273.

[81] WOOLFE K F. A scaled physical model for underwater sound radiation from a partially submerged cylindrical shell under impact [D]. Georgia Institute of Technology,2012.

[82] 尾形克彦. 现代控制工程[M]. 卢伯英,佟明安,译. 5版. 北京:电子工业出版社,2012.

[83] KIM H,SOHN J,JEON J,et al. Reduction of the radiating sound of a submerged finite cylindrical shell structure by active vibration control[J]. Sensors (14248220),2013,13(2):2131-2147.

[84] KIRCALI O F,YAMAN Y,NALBANTOGLU V,et al. Active Vibration Control of a Smart Beam by Using a Spatial Approach[M]// New Developments in Robotics Automation and Control. 2008:1318-1322.

[85] Kuhn G F,Morfey C L. Transmission of low-frequency internal sound through pipe walls [J]. Journal of Sound and Vibration,1976,47(2):147-161.

[86] KUMAR G V,RAJA S,SUDHA V. Finite element analysis and vibration control of a deep composite cylindrical shell using MFC actuators[J]. Smart Materials Research,2012,2012 (2090-3561):123-136.

[87] KUMAR P,JANGID R S,REDDY G R. Response of piping system with semi-active variable stiffness damper under tri-directional seismic excitation[J]. International Journal of Structural Engineering,2013,258(2):130-143.

[88] CHENG L. Fluid-structural coupling of a plate-ended cylindrical shell: vibration and internal sound field[J]. Journal of Sound & Vibration,1994,174(5):641-654.

[89] LAPLANTE W. Active control of vibration and noise radiation from fluid-loaded cylinder using active constrained layer damping [J]. Journal of Vibration and Control,2002,8(6):877-902.

[90] LEISSA A W. Vibration of Shells [M]. NASA,1973.

[91] LENIOWSKA L,MAZAN D. MFC sensors and actuators in active vibration control of the circular plate[J]. Archives of Acoustics,2015,40(2):257-265.

[92] LESMEZ M W,WIGGERT D C,HATFIELD F J. Modal analysis of vibrations in liquid-filled piping systems[J]. Journal of Fluids Engineering,1990,112(3):311-318.

[93] LESMEZ M W,WIGGERT D C,HATFIELD F J. Modal analysis of vibrations in liquid-filled piping systems[J]. Journal of Fluids Engineering,1990,112(3):311-318.

[94] LI J. Positive position feedback control for high-amplitude vibration of a flexible beam to a principal resonance excitation[J]. Shock and vibration,2010,17(2):187-204.

[95] LI D S,CHENG L,GOSSELIN C M. Optimal design of PZT actuators in active structural acoustic control of a cylindrical shell with a floor partition [J]. Journal of Sound &

Vibration,2004,269(3 -5):569 -588.

[96] LIEW K M,HE X Q,KITIPORNCHAI S. Finite element method for the feedback control of FGM shells in the frequency domain via piezoelectric sensors and actuators[J]. Computer Methods in Applied Mechanics & Engineering,2004,193(3/5):257 -273.

[97] SWINBANKS M A,DALEY S. Advanced submarine technology – project M control theory report. phase 1[J]. Advanced Submarine Technology Project M Control Theory Report Phase,1993.

[98] JOHNSON M E,ELLIOTT S J. Active control of sound radiation from vibrating surfaces using arrays of discrete actuators[J]. Journal of Sound and Vibration,1997,207(5):743 -759.

[99] CHANG M I J,SOONG T T. Optimal controller placement in modal control of complex systems[J]. Journal of Mathematical Analysis and Applications,1980,75(2):340 -358.

[100] TOKHI M O,MAMOUR K,HOSSAIN M A. Adaptive active noise and vibration control [C]// International Conference on Control. IET,2002.

[101] TOKHI M O,LEITCH R R. Active Noise Control [M]. clarendon,1992.

[102] XU M B,SONG G. Adaptive control of vibration wave propagation in cylindrical shells using SMA wall joint [J]. Journal of Sound and Vibration,2004,278(1):307 -326.

[103] FARES M E,YOUSSIF Y G,ALAMIR A E. Minimization of the dynamic response of composite laminated doubly curved shells using design and control optimization [J]. Composite Structures,2003,59(3):369 -383.

[104] MAILLARD J P,LAGÖ T L,FULLER C R. Fluid wave actuator for the active control of hydraulic pulsations in piping systems[J]. Proceedings of SPIE – The International Society for Optical Engineering,1999,3727:1806 -1812.

[105] BODSON M. adaptive Algorithms for active noise and vibration control [R]. Final Progress Report – Army Research Office,2000.

[106] MARTINO,AJANGNAY O A. Hybrid time – frequency domain adaptive filtering algorithm for electrodynamic shaker control [J]. Journal of Engineering and Computer Innovations, 2011,2(10):191 -205.

[107] RUZZENE M,BAZ A. Active/passive control of sound radiation and power flow in fluid – loaded shells[J]. Thin – Walled Structures,2000,38(1):17 -42.

[108] WINBERG M,JOHANSSON S,L HÅKANSSON,et al. Active vibration isolation in ships:a pre – analysis of sound and vibration problems[J]. International Journal of Acoustics and Vibrations,2005,10(4):175 -196.

[109] CARESTA M. Active control of sound radiated by a submarine in bending vibration[J]. Journal of Sound and Vibration,2011,330(4):615 -624.

[110] CARESTA M,KESSISSOGLOU N J. Acoustic signature of a submarine hull under harmonic excitation[J]. Applied Acoustics,2010,71(1):17 -31.

[111] CARESTA M,KESSISSOGLOU N J. Structural and acoustic responses of a fluid – loaded

cylindrical hull with structural discontinuities[J]. Applied Acoustics,2009,70(7):954-963.

[112] CARESTA M,KESSISSOGLOU N. Active control of sound radiated by a submarine hull in axisymmetric vibration using inertial actuators[J]. Journal of Vibration and Acoustics, 2012,81(134):1-8.

[113] CARESTA M,KESSISSOGLOU N J. Purely axial vibration of thin cylindrical shells with shear-diaphragm boundary conditions[J]. Applied Acoustics,2009,70(8):1081-1086.

[114] MAZUR, KRZYSZTOF, PAWELCZYK, et al. Internal model control for a light-weight active noise-reducing casing [J]. archives of acoustics,2016,41(2):315-322.

[115] MERZ S,KESSISSOGLOU N,KINNS R,et al. Minimisation of the sound power radiated by a submarine through optimisation of its resonance changer[J]. Journal of Sound and Vibration,2010,329(8):980-993.

[116] MERZ S,KINNS R,KESSISSOGLOU N. Structural and acoustic responses of a submarine hull due to propeller forces[J]. Journal of Sound and Vibration,2009,325(1-2):266-286.

[117] COHEN M. Classical Mechanics:a Critical Introduction [M]. The Pennsylvania State University CiteSeerX Archives,2012.

[118] LI M,LIM T C,SHEPARD W S. Modeling active vibration control of a geared rotor system [J]. Smart Materials & Structures,2004,13(3):449.

[119] QATU M S,SULLIVAN R W,WANG W. Recent research advances on the dynamic analysis of composite shells:2000-2009 [J]. Composite Structures,2010,93(1):14-31.

[120] MONNER H P,MONNER H P. Smart materials for active noise and vibration reduction [C]. Noise and Vibrations-Emering Methods. 2005.

[121] MOUSSOU P. An excitation spectrum criterion for the vibration-induced fatigue of small bore pipes[J]. Journal of Fluids & Structures,2003,18(2):149-163.

[122] NADER JALILI. Piezoelectric-Based Vibration Control:From Macro to Micro/Nano Scale Systmes [M]. Springer,2013.

[123] UR R N,NAUSHAD A M. Active vibration control of a piezoelectric beam using PID controller:Experimental study[J]. Latin American Journal of Solids and Structures,2012, 9(6):657-673.

[124] TRAJKOV N T,KÖPPE H,GABBERT U. Vibration control of a funnel-shaped shell structure with distributed piezoelectric actuators and sensors [J]. Smart Materials & Structures,2006,15(4):1119-1132.

[125] OLSON D E. Pipe vibration resting and analysis[J]. Cement & Concrete Research,1999, 29(7):1099-1102.

[126] ONODA J,HANAWA Y. Actuator placement optimization by genetic and improved simulated annealing algorithms [J]. AIAA Journal,1993,31:1167-1169.

[127] GARDONIO P,FERGUSON N S,FAHY F J. A modal expansion analysis of noise transmission through circular cylindrical shell structures with blocking masses[J]. Journal of Sound & Vibration,2001,244(2):259-297.

[128] LUEG P. Process of silencing sound oscillations［P］. Patent US 2,043,416,filed:March 8,1934,patented:June 9,1936.

[129] LUEG P. Verfahren zur dämpfung von schallschwingungen［P］. German Patent No. 655 508,filed:Jan. 27,1993,patented:Dec. 30,1937.

[130] MIN P,HILLIS A,JOHNSTON N. Active control of fluid－bome noise in hydraulic systems using in－series and by－pass structures［C］// Ukacc International Conference on Control. 2014:355－360.

[131] PAN X,YAN T,JUNIPER R. Active control of low－frequency hull－radiated noise［J］. Journal of Sound & Vibration,2008,313(1－2):29－45.

[132] HATCH M R. Vibration simulation using MATLAB and ANSYS［M］. Chapman & Hall/CRC,2000.

[133] LOCHMATTER P,KOVACS G,ERMANNI P. Development of a shell－like electroactive polymer (EAP) actuator［D］. Swiss Federal Institute of Technology Zurich,2007.

[134] PHOHOMSIRI P,UDWADIA F E,BREMEN H V. Time－delayed positive velocity feedback control design for active control of structures［C］// American Society of Civil Engineers. American Society of Civil Engineers,2006:690－703.

[135] MAO Q B,PIETRZKO S. Control of Noise and Structural Vibration:A Matalb－Based Approach［M］. Springer,2013.

[136] URALAM R N,NAUSHAD A M. Active vibration control of a piezoelectric beam using PID controller:experimental study［J］. Latin American Journal of Solids & Structures,2012,9(6):657－673.

[137] SAMBAVEKAR R V,PATEL S J,PATHARE H S,et al. Active vibration control of a cantilever beam using PZT PATCH (SP－5H)［J］. International Journal of Engineering and Technical Research,2015,3(5),37－39.

[138] DALEY S,JOHNSON F A,PEARSON J B,et al. Active vibration control for marine applications［J］. Control Engineering Practice,2004,12(4):465－474.

[139] DALEY S,JOHNSON F A. the "smart spring" mounting system:a new active control approach for isolating marine machinery vibration［J］. Elsevier Ifac Publications,2004,37(10):641－646.

[140] KELLY S G. 机械振动［M］. 贾启芬,刘习军,译. 北京:科学出版社2002.

[141] SHIELDS W H,RO J J,BAZ A M. Control of sound radiation from a plate into an acoustic cavity using active piezoelectric-damping composites［J］. Smart Materials & Structures,1999,7(1):1.

[142] AHN S S,RUZZENE M. Optimal design of cylindrical shells for enhanced buckling stability:Application to supercavitating underwater vehicles［J］. Finite Elements in Analysis and Design,2006,42(11):967－976.

[143] ELLIOTT S J. Signal Processing for Active Control［M］. Elsevier Ltd,2001.

[144] MITRA S K. Digital signal processing:a computer－based approach［M］. 2版. 北京:清

华大学出版社,2006.

[145] MERZ S,KESSISSOGLOU N,KINNS R,et al. Reduction of the sound power radiated by a submarine using passive and active vibration control[J]. Proceedings of acoustics,2009, Australia.

[146] KUO S M. Design of Active Noise Control Systems With the TMS320 Family [R]. Application note,1996.

[147] KUO S M,MORGAN D R. Active Noise Control Systems:Algorithms and DSP Implementations [M]. WILEY,1996.

[148] MOON S H. Finite element analysis and design of control system with feedback output using piezoelectric sensor/actuator for panel flutter suppression [J]. Finite Elements in Analysis and Design,2006,42(12):1071-1078.

[149] HASHEMINEJAD,SEYYED M,RABBANI,et al. Active transient sound radiation control from a smart piezocomposite hollow cylinder [J]. Archives Of Acoustics,2015,40(3): 359-381.

[150] SILCOX R J,ELLIOTT S J. Active control of multi-dimensional random sound in ducts [R]. NASA,1990.

[151] SINGIRESU S R. 机械振动[M]. 4 版. 李欣业,张明路,译. 北京:清华大学出版社,2009.

[152] SMITH P W. Phase velocities and displacement characteristics of free waves in a thin cylindrical shell [J]. The Journal of the Acoustical Society of America,1955,27(6):1065.

[153] SONG Y,WEN J,YU D,et al. Reduction of vibration and noise radiation of an underwater vehicle due to propeller forces using periodically layered isolators[J]. Journal of Sound and Vibration,2014,333(14):3031-3043.

[154] ŠTEFAN MARKUŠ. The Mechanics of Vibrations of Cylindrical Shell [M]. Elsevier,1988.

[155] STEPHEN ELLIOTT. 主动控制中的信号处理[M]. 翁震平,等,译. 北京:国防工业出版社,2014.

[156] LEON S L. 线性代数[M]. 张文博,张丽静,译. 北京:机械工业出版社,2010.

[157] KARRIS S T. Signals and systems with matlab applications [M]. 2nd edition. Orchard Publications,2003.

[158] TANG Y Y,SILCOX R J,ROBINSON J H. Sound transmission through cylindrical shell structures excited by boundary layer pressure fluctuations [R]. NASA Langley Technical Report Server,2013.

[159] KIM T,IVANTYSYNOVA M. Active vibration control of axial piston machine using higher harmonic least mean square control of swash plate[J]. 10th International Fluid Power Conference,2016.

[160] GUO T,LIU Z,CAI L. An improved force feedback control algorithm for active tendons [J]. Sensors,2012,12(8):11360-11371.

[161] TSAHALIS D T, KATSIKAS S K, MANOLAS D A. A genetic algorithm for optimal positioning of actuators in active noise control: results from the ASANCA project[J]. Aircraft Engineering and Aerospace Technology, 2000, 72(3):252-258.

[162] TSOUVALAS A, METRIKINE A. Structure-borne wave radiation by impact and vibratory piling in offshore installations: from sound prediction to auditory damage[J]. Journal of Marine Science and Engineering, 2016, 4(3):44-69.

[163] VARIYART W, BRENNAN M J. Active control of the n = 2 axial propagating wave in an infinite in vacuo pipe[J]. Smart Materials & Structures, 2004, 13(1):126-133(8).

[164] CARIGNAN C R, VANDERVELDE W E. Number and placement of control system components considering possible failures[C]// American Control Conference. IEEE, 1984(7):703-709.

[165] WACHEL J C, TISON J D. Vibrations in reciprocating machinery and piping systems[C]. Proceeding of the twenty-third tubro machinery symposium, 1987.

[166] WANG Z, SUN Y. Experimental research on active vibration control of pipe by inertial actuator and adaptive control[J]. Journal of Huaqiao University, 2014, 91(5):725-734.

[167] WEI L, JIE P. Measurement of sound-radiation from a torpedo-shaped structure subjected to an axial excitation [C]. Proceedings of 20th International Congress on Acoustics, ICA 2010, Sydney, Australia.

[168] JOHNSON W R, ASLANI P, HENDRICKS D R. Acoustic radiation mode shapes for control of plates and shells [C]. ICA 2013, Montreal, Canada.

[169] WILLIAMS R B, PARK G, INMAN D J, et al. An overview of composite actuators with piezoceramic fibers [J]. Proceedings of SPIE-The International Society for Optical Engineering, 2002, 4753:421-427.

[170] GAO X, CHEN Q. Active vibration control for a bilinear system with nonlinear velocity time-delayed feedback[J]. Lecture Notes in Engineering & Computer Science, 2013, 2206(1):2037-2042.

[171] PAN X, HANSEN C H. Active control of vibration transmission in a cylindrical shell [J]. Journal of Sound and Vibration, 1997, 203(3):409-434.

[172] PAN X, YAN T, JUNIPER R. Active control of low-frequency hull-radiated noise[J]. Journal of Sound and Vibration, 2008, 313(1-2):29-45.

[173] PAN X, FORREST J A, JUNIPER R G. Optimal design of a control actuator for sound attenuation in a piping system excited by a positive displacement pump [C]. Proceedings of acoustics, 2009.

[174] XU M. Adaptive-passive and active control of vibration and wave propagation in cylindrical shells using smart materials [D]. The Graduate Faculty of The University of Akron, 2005.

[175] YAN J, LI T Y, LIU J X, et al. Input power flow in a submerged infinite cylindrical shell with doubly periodic supports[J]. Applied Acoustics, 2008, 69(8):681-690.

[176] HUANG Y M, CHEN C C. Optimal design of dynamic absorbers on vibration and noise

control of the fuselage[J]. Computers and Structures,2000,76(6):691-702.
[177] YOUSEFI A. active vibration control of smart structures using piezoelements. [C]// CanSmart Workshop. 1998.
[178] YUAN H G,JR W,LIM T C,et al. Experimental analysis of an active vibration control system for gearboxes[J]. Smart Materials and Structures,2004,13(5):1230.
[179] YING Z G,NI Y Q,KO J M. Semi-active optimal control of linearized systems with multi-degree of freedom and application[J]. Journal of Sound and Vibration,2005,279(1-2): 373-388.
[180] ZHANG K,SCORLETTI G,ICHCHOU M N,et al. Robust active vibration control of piezoelectric flexible structures using deterministic and probabilistic analysis[J]. Journal of Intelligent Material Systems & Structures,2013,25(6):665-679.
[181] ZHANG X M,LIU G R,LAM K Y. Vibration analysis of thin cylindrical shells using wave propagation approach[J]. Journal of Sound and Vibration,2001,239(3):397-403.
[182] ZHAO G. Active structural acoustic control of rotating machinery using piezo-based rotating inertial actuators[C]// Eccomas Thematic Conference on Smart Structures & Materials Smart. 2015.
[183] LIU Z,YI C. The analysis of vibro-Acoustic coupled characteristics of ball mill cylinder under impact excitation [J]. Modern Applied Science,2008,2(6),37-40.
[184] ZORIC N,SIMONOVIC A,MITROVIC Z,et al. Active vibration control of smart composite beams using PSO-optimized self-tuning fuzzy logic controller[J]. Journal of Theoretical and Applied Mechanics,2013,51(2):275-286.
[185] 陈斌.浮筏隔振系统建模及振动主动控制研究[D].合肥:中国科学技术大学,2008.
[186] 陈斌,李嘉全,邵长星,等.浮筏多通道协调振动主动控制实验研究[J].实验力学, 2008,23(3):248-254.
[187] 陈仁文,孙亚飞,熊克,等.飞机座舱结构声的弹性波主动控制研究[J].南京航空航天大学学报,2003,35(5):489-493.
[188] 单树军,何琳.可控阻尼半主动冲击隔离技术研究[J],2006,25(5):144-147.
[189] 丁吉,姜涛,蒋宁.基于悬臂梁振动的主动控制技术研究[J].长春工业大学学报(自然科学版),2007,28(3):312-315.
[190] 董维杰,贾艳丽,杨明刚.基于分时复用压电自感知执行器的悬臂梁振动主动控制研究[J].振动与冲击,2009,28(10):150-153.
[191] 董兴建,孟光.压电悬臂梁的动力学建模与主动控制[J].振动与冲击,2005,24(6): 54-56.
[192] 方同,薛璞.振动理论及应用[M].西安:西北工业大学出版社,2004.
[193] 缑新科,付兆祥.含压电片悬臂梁的LQR振动主动控制[J].机械与电子,2009(11): 20-22.
[194] 谷宁昌,张煜盛.隔振器布置和安装研究[J].船舶工程,2007,29(2):30-33.
[195] 顾俭,姜哲.任意边界梁的声辐射模态伴随系数测量[J].振动工程学报,2004,17

(1):86-90.

[196] 关守平,尤富强,徐林,等.计算机控制:理论与设计[M].机械工业出版社,2012.

[197] 和卫平,陈美霞,高菊,等.基于统计能量法的环肋圆柱壳中、高频振动与声辐射性能数值分析[J],中国舰船研究,2008,3(6),7-12.

[198] 贺学锋,印显方,杜志刚,等.悬臂梁压电振动能采集器的集总参数模型和实验验证[J].纳米技术与精密工程,2012,10(2):108-110.

[199] 胡红生,钱林方.移动质量激励悬臂梁振动主动控制系统[J].传感器技术,2005,24(9):49-53.

[200] 胡寿松.自动控制原理[M].5版.北京:科学出版社,2007.

[201] 姜荣俊,何琳,JIANG R J,等.有源振动噪声控制技术在潜艇中的应用研究[J].噪声与振动控制,2005,25(2):1-6.

[202] 居余马,等.线性代数[M].2版.北京:清华大学出版社,2002.

[203] 李嘉全.浮筏系统的振动主动控制技术研究[D].合肥:中国科学技术大学,2008.

[204] 李克安.等截面悬臂梁的振动分析[J].湖南大学邵阳分校学报,1989,2(1):18-21.

[205] 李庆扬,王能超,李大义.数值分析[M].5版.北京:清华大学出版社,2008.

[206] 李维嘉,曹青松.船舶振动主动控制的研究进展与评述[J].中国造船,2007,48(2):68-79.

[207] 李再承,侯国祥,吴崇建.管系湍流噪声辐射研究方法进展[J].中国舰船研究,2007,2(1):34-38.

[208] 李志斌,陈红,钟宇明,等.压电陶瓷驱动的柔性悬臂梁PPF振动控制实验研究[J].深圳职业技术学院学报,2010,09(5):1-5.

[209] 梁青,段小帅,陈绍青,等.基于滤波x-LMS算法的磁悬浮隔振器控制研究[J].振动与冲击,2010,29(7):201-204.

[210] 梁旭,黄明.现代智能优化混合算法及其应用[M].北京:电子工业出版社,2012。

[211] 鲁力.基于动磁式主动吸振器的振动主动控制[D].合肥:中国科学技术大学,2007.

[212] 陆轶,顾仲权.直升机结构响应主动控制作动器优化设计研究[J].振动与冲击,2007,26(3):23-26.

[213] 牛军川,宋孔杰.船载柴油机浮筏隔振系统的主动控制策略研究[J].内燃机学报,2004,22(3):252-256.

[214] 庞中华,崔红.系统辨识与自适应控制:MATLAB仿真[M].北京:北京航空航天大学出版社,2009.

[215] 钱学森,宋健.工程控制论(上册)[M].北京:科学出版社,1980.

[216] 阮宏博.基于遗传算法的工程多目标优化研究[D].大连:大连理工大学,2007.

[217] 沈少萍.特征模型自适应控制方法在悬臂梁振动主动控制中的应用[J].振动与冲击,2007,26(11):115-119.

[218] 师汉民,黄其柏.机械振动系统-分析、建模、测试、对策(上册)[M].武汉:华中科技大学出版社,2013.

[219] 史蒂芬·埃利奥.主动控制中的信号处理[M].翁震平,等,译.北京:国防工业出版

社,2014.
- [220] 史荣昌,魏丰.矩阵分析(第三版)[M].北京:北京理工大学出版社,2010.
- [221] 宋玉超,于洪亮.隔振器最佳选择方案[J].大连海事大学学报,200,33(1):87-89.
- [222] 同济大学数学系.工程数学:线性代数[M].5版.北京:高等教育出版社,2007.
- [223] 王波,殷学纲,黄尚廉.压电自感知悬臂梁振动的主动控制研究[J].固体力学学报,2004,25(1):107-110.
- [224] 王二成,赵亚军,马晓雨.压电智能悬臂梁结构振动主动控制仿真[J].机械设计与制造,2011,(2):180-182.
- [225] 王建辉,顾树生.自动控制原理[M].北京:清华大学出版社,2007.
- [226] 王俊芳.自适应主动隔振的理论和实验研究[D].上海:上海交通大学,2008.
- [227] 闻邦椿.机械振动理论及应用[M].北京:高等教育出版社,2009.
- [228] 吴正国,尹为民,侯新国,等.高等数字信号处理[M].北京:机械工业出版社,2009.
- [229] 许本文,焦群英.机械振动与模态分析基础[M].北京:机械工业出版社,1998.
- [230] 严济宽,沈密群.怎样选用隔振器[J].噪声与振动控制,1982(06):1-7.
- [231] 杨铁军,靳国永,李玩幽,等.舰船动力装置振动主动控制技术研究[J].舰船科学技术,2006,28(2):46-53.
- [232] 姚熊亮,顾玉钢,杨志国.压电类智能结构在船体振动控制方面的应用研究[J].哈尔滨工程大学学报,2004,25(6):695-699.
- [233] 于哲峰.压电堆作动器微振动控制平台控制效果评估与控制力效率分析[J].噪声与振动控制,2009,29(6):170-174.
- [234] 张磊,束立红,何琳,等.磁致伸缩作动器的设计与性能分析[J],海军工程大学学报,2008,18(4):75-79.
- [235] 张嗣瀛,高立群.现代控制理论[M].北京:清华大学出版社,2006.
- [236] 张志涌,等.精通MATLAB R2011a[M].北京:航空航天大学出版社,2011.
- [237] 赵冬冬,张京军,王二成.柔性悬臂梁振动主动控制研究[J].山西建筑,2007,33(18):3-4.
- [238] 郑世杰,王晓雪.基于固体壳单元的功能梯度材料板壳主动控制模拟仿真[J].航空动力学报,2006,21(6):1075-1079.
- [239] 周星德,陈道政,杜成斌.框架结构主动控制最优时滞研究[J].计算力学学报,2007,24(2):246-249.
- [240] 周英,王晓雷,郑钢铁.空气弹簧隔振器主动控制的鲁棒控制方法研究[J].振动与冲击,2007,26(1):125-129.
- [241] 朱灯林,俞洁.压电智能悬臂梁的主动振动控制[J].河海大学常州分校学报,2005,19(4):1-4.
- [242] 朱石坚,楼京俊,何其伟,等.振动理论与隔振技术[M].北京:国防工业出版社,2006.
- [243] KORUK H, DREYER J T, SINGH R. Modal analysis of thin cylindrical shells with cardboard liners and estimation of loss factors[J]. Mechanical Systems and Signal Processing, 2014, 45(2):346-359.